Nuts And Bolts

A Guide to Software Engineering
in a world of robots,
space ships
and prosthetic brains

Trevy Burgess

Nuts And Bolts

A Guide to Software Engineering

in a world of robots, space ships and prosthetic brains

Trevy Burgess

Copyright © 2018 Trevy Burgess

First Edition

Published by Trevy Burgess

ISBN: 978-0-359-14777-9

Dedicated to my parents...
They always believed in me –
even when they were scolding me.

Table of Contents

Intuition is absolutely not something natural.
It is something gleamed from an enormous
amount of practice and experience.
-- Toriko (Anime), Episode 77 --

Contents

Nuts And Bolts
A Guide to Software Engineering

Parable:

One day three monks walked down a path.

The First said to the Second,
"What would you do if you saw the Buddha?"

The Second said,
"I would prostrate before him and ask him how best to solve
my software engineering dilemma."

They walked on...

The Second then asked the First,
"What would you do?"

The First said,
"I would spit in his face and kick his ass,
for giving us confusing advice."

They walked on...

Both monks looked at the Third.

The Third said,
"There is no Buddha. There is only you and I,
trying our best
to help our customers with their IT needs."

1. INTRODUCTION

Scientists investigate that which already is;
Engineers create that which has never been.
-- Albert Einstein --

E ngineering – The meticulous use of scientific principles, personal experience, human understanding, and a dash of hocus-pocus, to bring about wonders that serve the human need.

Software engineering is just like any other engineering endeavor. Mistakes can and do cost lives and destroy livelihoods.

Engineering is based on the discovery and rediscovery of principles countless people have worked hard to discover. In software engineering, these principles are known as design patterns. Design patterns express themselves through the three aspects to Software Engineering:

- Project Management
- Feature Management
- Feature Development

Product Life Cycle Snapshot

Comment: People in the information technology field constantly strive to be recognized as an engineering discipline. That is impossible since traditional engineering is based on the application of physical laws to solve everyday problems.

Information technology doesn't obey physical laws, but informational laws.

> **Comment:** One day people will realize that the universe is fundamentally an informational system and not actually physical. At that point the various engineering disciplines will be seen for what they are — as various branches of information technology. Even now engineers are more informational technology professionals than nuts-and-bolts people.

1.0.1. **The need**
Before we begin a project, all we have is a vaguely defined set of needs:

- I want to track how many customers buy sprockets every hour.
- I want to track how our sprockets are being used.
- I want to ensure the highest quality sprockets, using the cheapest processes possible.

1.0.2. **The challenge**
It is not enough to know what we want.

We must define:
- Steps needed to achieve the goal
- Resources available
- Success criteria

Requirements
Gathering

1.0.3. **The team**
The purpose of the team is to take the vaguely defined requests of the client and create a solution that fulfills the client's needs.

All members of the team are involved in defining implementation details. It starts with defining requirements and ends with a feature the client is satisfied with, and a list of new features the client wants.

1.0.4. **The Three Parts**
The three parts of the development team work in concert to achieve the goal of customer satisfaction.

Think of the team as the three legs of a stool. The teams are: *1, A,* and α. The cross bracing is the collaboration skills used to ensure a unified goal. The seat is the where the unity of the team comes together to service the customer.

Just like any stool, all parts of the team must work in unity to allow for maximum customer satisfaction.

Leg 1 – Feature Management

The function of feature management is to collect the requirements and implement processes that verify what is developed is what is needed.

The traditional name for this team is **testers**. However I don't like that name. That implies all they do is look for bugs and slow development down.

The real power of this team is to:

1. Clearly document the expectations of the currently implemented features.
2. Discover and document all hidden assumptions.
3. Ensure new features don't change existing functionality.
4. Ensure routine refactoring exercises don't change existing functionality.
5. Document a unified vision of the product that conforms to customer expectations.
6. Give customers confidence that the product will satisfy their needs.

As you can see, feature management is more than catching bugs. It ensures we deliver a quality product designed to exceed the expectations of the customer.

Leg A - Feature Development

In feature development, we take the features we want to develop from a previously generated list of work items (backlog) and develop them.

We then use the verification processes developed by Team 1 to make sure that what we have is what our stakeholders have specified.

Leg α - Project Management

Project management controls the entire process. Project Managers make sure there is continuous and open communication between all stakeholders.

The process is as follows:

1. We define what feature we need

The project manager sits down with the feature management team, feature development team, customers and others. They figure out what features need to be developed over the next sprint (development cycle). The features are then added to a weighted list (backlog).

One important thing about the list is that the requested features must be doable in less than one sprint cycle (typically 2 weeks).

If it is not a simple addition of functionality, then it's an **epic.** Epics need team collaboration to develop, and its own set of features to implement.

It doesn't matter if we make mistakes, since mistakes and confusion are an unavoidable part of the development cycle.

The main problem with sprint planning is that costing feature development is hard. A simple feature can be impossible, if the basic infrastructure doesn't support it. On the other hand, some hard-seeming features only require ten minutes to implement.

2.   We define the acceptance criteria
The feature management team creates the processes needed to verify that the feature developed is in fact what was requested.

3.   We develop the feature
The feature development team works on the features requested by the customers.

4.   We verify the correctness of the newly developed functionality
This is where the processes developed by the feature management team come into play. The (ideally) automated test cases are used to ensure the new feature works as intended and existing functionality hasn't changed its behavior.

5.   At the end of the cycle, we review our progress
The feature management team, the feature development team and the product management team get together to discuss what went well, what went wrong, and what could be improved.

6.   We restart the process, learning from past mistakes
We are building functionality incrementally, so that the product doesn't make unexpected detours. This is like marking your trail as you walk, instead of running full speed ahead. It is slower, but when mistakes happen, they are faster and easier to fix.

 Note: Although we talk about three teams with separate responsibilities, in reality everyone should be involved in all aspects of development.

In fact, switching roles on a regular basis is becoming popular.

Comment: Waterfall has a bad rap. However it is fundamental to the development cycle. First, it defines the overall theme of the solution. Second, it defines the epics of the product and feature sets. Once defined, agile then fills in the blanks, by working on multiple required features at the same time.

1.1. <u>**About This Book**</u>

Obscurity and obsessive abstraction
are two of the worst problems
that affect software development;
they combine into a form of willful ignorance
that makes us write crappy code.
-- Marco Tabini --

My decade+ years of experience as a software developer center on MS Windows programming, the C# programming language, the MS.NET framework, and the Visual Studios suite of development products. As a result, this book will use these technologies to exemplify the principles embodied in this book.

1.1.1. <u>Sample Code and Blog</u>
Here is some of my published work...

<u>Sample Code</u>
*   Source: https://github.com/TrevyBurgess

<u>UWP App - Zee Events Manager</u>
*   https://www.microsoft.com/en-us/p/zee-events-manager/9nblggh4vb78

<u>Blog</u>
*   http://trevy-korner.blogspot.com/

1.1.2. <u>Pictures and Stories</u>
I don't believe it is possible to learn complex principles without concrete examples. I believe the best way to learn is through stories and pictures. As a result, I will be illustrating the various concepts using stories.

I will not be using the Unified Modeling Language (UML), since I don't believe it has practical use in learning new concepts.

Feel free to port the principles to your favorite programming language and development framework, if you find them useful.

You may also borrow my source code if you find it useful.

This is a work in progress book. I will be updating it as my understanding of software development evolves. I look forward to your feedback and hope this book is useful to you.

1.1.3. <u>Images used in this Book</u>
All images come from Microsoft's open image collection.

1.1.4. <u>Conventions used in this book</u>
The names of design patterns are marked with braces, like {Model-View-Controller}.

Also, various comments are sprinkled through the book. They call out to key points, to allow for skimming of the book.

	Reference: All references are marked with this symbol.
	Definition: Used to define a concept or idea.
	Comment: A commentary on the main subject.
	Note: Something to be aware of that may or may not be important.
	Take Home: The main point of the explanation.
	Warning: This refers to anything that can cause a bug or cause data to be lost.

<u>Intended Audience</u>
This book is aimed at medium to advanced software developers who are looking for a different take on the software development experience.

This book assumes the reader has taken basic training in software engineering. As a result, I might skip over some basics.

Also, this book assumes you are familiar with C#, the MS.NET framework, and Visual Studios. However, lack of this knowledge should not hamper you.

Above all, this book aims to give a different perspective on what you already know. I hope this satisfies your needs.

<u>Scope of the Book</u>
This book is intended to give an overview of the processes involved. With that overview, we can get more detailed information as needed.

Disclaimer

This is just my view on the design process. It is neither better nor worse than any other view. The important thing is that it be useful to you in your work.

Also, my examples revolve around C#, Visual Studios, and the .NET framework. That doesn't mean other frameworks or ways of doing things are either better or worst.

 Warning: This book contains some bad puns. Read at your own risk.

2. PROJECT MANAGEMENT

Software changes. This is a rather obvious statement, but it is a fact that must be ever present in the minds of developers and architects. Although we tend to think of software development as chiefly an engineering exercise, the analogy breaks down very quickly. When was the last time someone asked the designers of the Empire State building to add ten new floors at the bottom, put a pool on the top, and have all of this done before Monday morning?[1]

> I love deadlines.
> I like the whooshing sound they make as they fly by.
> -- Douglas Adams --

> Management is doing things right.
> Leadership is doing the right things.
> -- Peter Drucker --

Project management is the art of organizing all the diverse components of a project so that a product can be produced that satisfies the customer's needs, in a timely and cost-effect manner.

2.1. Project Manager

> Leaders must be close enough to relate to others,
> but far enough ahead to motivate them.
> -- John C Maxwell --

Project managers define the atmosphere team members work in. They have higher level knowledge of requirements since they work with upper management. They are the one responsible for the success or failure of a project. Finally, they have access to the resources the team needs to succeed. As such, project managers are a key resource in making the team a success.

2.1.1. Project Manager as Knowledge Broker

One fundamental task the project manager has is keeping track of the skills of everyone on the development team. This information should be made available to the entire team, preferably in the form of a WIKI.

In addition, the project manager needs to make sure that every team member is actively maintaining their area of the (internal) team WIKI.

---

[1] http://msdn.microsoft.com/en-us/library/ee817667.aspx

Without the team actively updating the WIKI, much knowledge will be lost when the team member moves to their next project.

 Take Home: Make updating the WIKI a recurring work item for all team members. This will ensure proper knowledge retention, when change happens.

2.1.2. <u>Project Manager as Coach</u>

There is a widespread belief that you should hire the very best people available before starting a project.

There are several flaws in that thinking:

1. The current hiring practices are so bad that many times the best people are passed over for a mediocre person who is good at interviews.
2. It can take weeks or months of looking before a suitable candidate is found. Since time is money, that represents thousands or millions of dollars worth of lost revenue.
3. It takes time for new people to ramp up on an existing project, even if they are as smart as Einstein.
4. Knowing a specific technology will not grantee success. Not being an expert on a technology will not guarantee failure.

The tech world is changing so fast that cutting-edge technologies can become obsolete in a blink of an eye.

The best solution is to get people who are trainable and then train them. The ideal person has a demonstrated ability to learn.

The best advice I can give here is to get the book, *How to be a STAR at Work*, by Robert E. Kelley. That book is, in my opinion, one of the greatest books out there on productivity. This book is not just for managers, but everyone who wants to be successful.

Comment: In my opinion, the Contract-to-Hire route of obtaining fulltime help is the best, since you can try them out before hiring, and you save precious time, instead of wasting it on interviews. Also you know what you're getting before you make a full-time commitment.

 Reference: *How to be a STAR at Work*, by Robert E. Kelley.
ISBN-10: 0812931696
ISBN-13: 978-0812931693

2.2. Scrum

Scrum is like your mother-in-law,
it points out ALL your faults.
-- Ken Schwaber --

The purpose of scrum is to organize the day to day activities of a team, ensuring everyone is on the same page as to what needs to be done.

The sprint is the basic timeframe for doing work. The standard is two weeks. However some prefer either one-week sprints, three-week sprints, or one-month sprints.

2.2.1. Scrum Master

The role of the scrum master is to coordinate activities within a sprint and act as the go-between for the team and management.

Various tools such as Jira and TFS aid their work.

Interruptions

From research on how humans operate, we know that it takes time and effort to adjust to any role we are assigned to.

Also, when you start a project, it takes time to go into the flow of things before becoming productive. This is especially true of large projects.

Every time there is an interruption, we need to start anew, wasting hours of time before real work can begin.

Changing Requirements

Another big reason for failure is when managers make requests on the team at the spur of the moment. This causes team members to run around like dancing chickens.

Then managers wonder why team members aren't doing their jobs properly, or why deadlines are slipping.

To avoid this problem, it is the job of the scrum master to protect the team from outside influence until the sprint is completed or cancelled. Of course, once the sprint is complete, the feature team has to show results.

 Note: It is the job of the Scrum Master to protect the team from interruptions, and interference from management.

2.2.2. <u>Planning Meeting</u>

At the beginning of the development cycle or sprint, representatives for all stakeholders get together and decide on the features that the development team needs to work on over the course of the sprint.

This committee creates an ordered list of items (backlog) for the development team to work on. This process should take no more than a day to complete and can involve two hour-long meetings. A break between sessions is necessary for the stakeholders to digest what happened in the first meeting and prepare for the next meeting.

Input from the development team is essential, since the time and cost involved in implementing a feature can vary greatly, depending on the current state of the product.

There are many ways to decide the priority of what is on the backlog and what is on the back burner. One possibility is giving all stakeholders a number of votes (Ex. 10). Then everyone places their votes on what they think is most important.

It is okay to put all your votes on one item. However, you will have no say on any other items. This way, priority is given to the work items that people feel add the most value.

Feature Verification
It is standard practice to write the feature code and the feature verification code at the same time, in order to save time.

This means that the members of the feature development team are working on their work items without the assurance that what they are doing is correct.

Only when both teams have finished can we verify that what was requested is what was delivered. As a result, next sprint has to focus on bug fixes.

One possible solution is to do test driven development. First we create tests everyone agrees captures the requirements. Then next sprint the feature is developed. This shouldn't be a problem if you're publishing once a month.

> **Comment:** First, let the feature management team fully define the feature as a set of tests. Next sprint, the feature development team implements the feature, using the tests implemented by the feature management team.

2.2.3. <u>Start of Sprint</u>

Once the backlog is finalized, it is time to work. But what should developers work on?

The traditional way is for developers to choose items at the beginning of a sprint. They then work on those items throughout the sprint.

The problem with this approach is that we need to decide at the beginning of the sprint what a developer can do to fill the time allotted. As a result, either too many items are chosen or too few items. Many times items need to be postponed, because they rely on undeveloped functionality.

Another problem is that it looks bad when developers don't finish items quickly or when they return items back to the backlog.

My preference is to only pull items into your own board when you are ready to work on a feature. I will assume this from now.

It is okay to push items back onto the backlog if it becomes too hard or time-consuming. This is preferable than just letting items sit on your plate.

 Warning: The backlog **must** be locked down between planning meetings. This will prevent pet projects from slipping in, which could negatively impact the product.

 Warning: Developers may only work on items on the backlog. Accepting drop-in requests from anyone is forbidden.

Warning: If a feature requires undeveloped functionality, then the feature must be returned to the backlog. It will be turned into an epic at the next sprint planning meeting.

2.2.4. <u>Morning Scrum Meetings</u>

The purpose of morning scrum meetings is to identify risks as they show up. For instance people can get stuck or go down the wrong path.

Every morning the team gets together and shares a 2-minute summary of progress. In-depth discussions are left for later.

Members would mention:
- Items that were taken from the backlog
- Items returned to the backlog.
- Issues encountered
- Any new ideas generated when implementing the item

Putting things back into the backlog is important. This allows the team to reevaluate priorities. Many times developers are forced to work on features that are unreasonable or too time-consuming. Returning items allow us to improve the quality of the work and the stability and morale of the team.

 Take Home: Putting things back into the backlog is important. This allows the team to reevaluate priorities.

 Comment: Those who don't have challenges to share and those who come late have to contribute to the weekly lottery or doughnut bucket.

2.2.5. <u>End of Sprint Retrospective</u>

 At the end of the sprint, everyone gets together and discuss what went well during the sprint, what went wrong, and what we learnt.

This is an essential part of the sprint since it sets the tone for the next sprint.

By its very nature, the software development process is an exploration into unknown territory. We need to know where we've been and what dangers we faced, so we can better face the dangers still ahead.

2.3. <u>Work Environment</u>

If people knew how hard I had to work to gain my mastery,
it would not seem so wonderful at all.
-- Michelangelo --

2.3.1. <u>Shared office Space</u>

The ideal work environment is where every member of the team works in the same room. This allows the potential for maximum communication. You just call, "What's the web site of, you know, where we had to put our stuff." Someone just turns around and answers.

Also, being in a shared room, we can quickly walk to the person and get advice.

Companies such as Apple have embraced this concept, to the consternation of its employees. Having a private office is always nice.

2.3.2. <u>Virtual Offices</u>
I believe the future will be virtual.

Imagine if you will, you put on VR goggles, etc. and enter a virtual reality room. All your colleges are milling around. In front of you are your keyboard and monitor(s).

People get together and work on a project. Then people move around, adjusting the environment as needed. In the real world, you are in your private office, or you could be at home.

 Reference: A vendor that works with virtual reality work and collaboration spaces is: https://altvr.com/

2.3.3. <u>Contingent Staff</u>
As a project manager, you will be faced with a major productivity block if you have contingent staff on your team.

Software companies hire contingent staff to offset fluctuating work needs. They can discard people as needed, or hire them as needed.

This means brand new people will enter your team. Being new code monkeys, they have three things that hinder their productivity.

1. The new vendor will not have full permissions. As a result, the first week or so will be wasted as they struggle to get unblocked.
2. They need to learn the code base. Having someone to mentor them is essential for rapid ramping up. An established on-boarding procedure is essential for your company.
3. They need to get familiar with who knows what. This is a major stumbling block, since this knowledge is usually in people's heads and not easily retrievable. This is why WIKIs are important.

Another important hindrance to productivity is that there is insufficient buy-in. Why is that?

1. Being outsiders, they may not be included in team meetings. Also, they are here today and gone tomorrow.

This hinders their ability to obtain all the information they need to do their jobs quickly and efficiently.

2. Vendors may not be included in daily scrum meetings or important discussions.

According to the scrum philosophy, everyone who is involved in a feature in development must be included. If this is not done, then you should expect delays as the vendor tries to figure out how to do something that half the team already knows how to do.

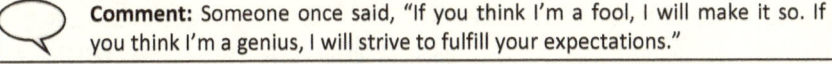 **Comment:** Without full buy-in from all parties, the product suffers from wasted talent and missed opportunities. "What's the point in me trying to contribute if no one cares?"

3. Brain drain

Many companies like having new people enter the team, since this represents new ideas. Unfortunately, when it comes to vendors, they are usually let go just as they come to a point where they can make a significant contribution.

What's the point in expressing innovative ideas, if you're not there to shepherd the idea to fruition?

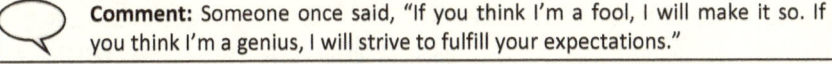 **Comment:** Someone once said, "If you think I'm a fool, I will make it so. If you think I'm a genius, I will strive to fulfill your expectations."

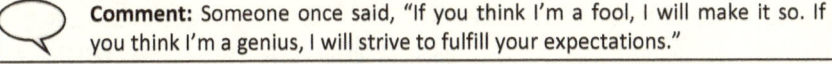 **Comment:** Some people hate their annual review. I say that is preferable to the alternative, which is being let go at fixed intervals or at a drop of a hat, regardless of performance.

3. ASPECTS OF BUSINESS SOLUTIONS

Simple systems are not feasible
because they require infinite testing.
-- Norman Ralph Augustine --

Before software can be reusable,
it first has to be usable.
-- Ralph Johnson --

When creating a product, we need to establish beforehand what we want, otherwise we won't get it. That of course comes from our backlog of items.

The question is, how do we make sure that what the product development team is working on is what is required?

The ideal way is to define the requirements, establish a process for verifying that the requirements have been met, and finally developing the feature in accordance with the requirements. This is Requirements Driven Development, sometimes referred to as Test Driven Development (TDD).

Cycling through this process advances our understanding of the requirements, which then results in input for our next sprint.

It's the job of the feature management team to make sure that what the feature development team produces is what is required.

So how do we verify the quality of the product?

We can't – not unless we understand the multiple aspects that make up a quality product.

When I was studying to be certified as a Microsoft Certified Solution Developer for the MS.NET Platform, I came across a book called *.Net Solution Architectures*. In it, they had the acronym 'PASS MADE'.

 Samples Code: https://github.com/TrevyBurgess

It stands for: **P**erformance, **A**vailability, **S**ecurity, **S**calability, **M**aintainability, **A**ccessibility, **D**eployability, and **E**xtensibility

Add **R**esponsiveness, **U**sability, **G**lobalization, and **S**ex-appeal, and we have – PASS-MADE-RUGS.

> **Comment:** People like comparing writing a program to building a house. If you use that analogy, then the source code is the blueprints for the house. The shipped application or package is the actual physical house.

It's time to walk through some of the aspects of enterprise solutions.

3.1. **P**erformance

If they try to rush me, I always say,
I've only got one other speed and it's slower.
-- Glenn Ford --

From the days since just after the dinosaurs invented the abacus to now, performance has always been considered important. However, putting too much emphasis on performance can be counter-productive.

Years ago, a company analyzed its software and discovered that it spent 96% of the time in one method. They spent weeks tweaking the method, but couldn't change the number. Later, they realized that they were tweaking a function that was meant to respond to user input. The limiting factor wasn't the method but the operator. The method was working correctly from the start.

People who obsess over performance are showing their Computer Age.
– Ha, ha, ha

Another drawback with over-emphasizing performance is a code base that is hard to maintain. Would anyone like some fries with that spaghetti code?

Let's **P.A.C.E.** through the things that pertain to performance...

3.1.1. **P**erformance Counters

Before we can consider optimizing for performance, we need to verify that there is a need. To do this, we need benchmarks to compare performance against. Benchmarks are typically collected through performance counters.

Performance counters keep track of everything from clock cycles to memory to network bandwidth. Each of these things affects not just performance, but the other quality matrices as well.

The list of common performance counters include:
- Memory Usage
- Network Bandwidth
- CPU cycles
- Object reference counts
- Thread usage
- Database connections

When dealing with performance, you need to ask the question "Is the customer happy?"

If the answer to that question is "Yes", then you should stop and move on to other areas. Trying to improve performance would be counter-productive and a waste of money, since no one will notice.

3.1.2. Algorithms

Programs are just the implementation of algorithms used to fulfill real-world needs.

As the saying goes, "Many roads lead to Rome". However, not all roads are easy to follow and some roads can be downright dangerous.

Speed-Maintainability Tradeoff

When selecting an algorithm, look for the simplest one available. This will allow for easier maintenance of the product.

Once the algorithm is selected and coded, we then check performance. The performance counters we use will differ depending on the intended use.

Some apps include:
- Single user app
- Game and graphics programs
- Server apps
- Stateless Web Service
- Machine Learning apps

If performance is acceptable, then we close the ticket.

When performance is too slow, then we look for more complex algorithms. Why is that? It's because faster algorithms have convoluted logic that may not be easy to follow or debug.

Remember, we never code in isolation. Someone else will take over once we leave the team.

Order of Operation
Another thing to worry about is the order of operation for an algorithm. (See p. 162)

Data Collections
Speed may not be important if the app only needs to perform the calculation once. With modern machines, these calculations complete in milliseconds.

However, if we are dealing with collections of data, speed may be an issue.

If the data items have no dependencies on each other, then we can use multithreading and stateless library functions to speed up the processing. Big data benefits from this the most.

Image and audio processing is a different problem, since the data points are related to each other. Then more complex algorithms become necessary.

3.1.3. Code Tweaking

NEVER use code tweaks to increase performance – unless performance analysis indicates that the code is affecting customer satisfaction.

There are two kinds of performance enhancements: Nonstandard code and best code practices.

Best code practices include using the most efficient algorithms available. It also includes using a consistent coding style. If everyone uses the same style, then departing from that style will make your code harder to read.

Of course you need to use your judgment when breaking the team's coding style. However you will need to clearly document why you needed to break the rule, so future developers don't accidently break the program.

Nonstandard practices include spaghetti code. It is okay to use spaghetti code in your project if:

1. Performance tests fail for the current implementation and no other mechanism is available.
2. You have buy-in from all parties
3. The code is properly documented

3.1.4. Environment

The environment consists of the operating system, necessary 3rd party applications, the network and hardware, to name a few variables.

Native code, such as C++ applications run directly on the hardware and use OS methods for standard operations. This code has to be recompiled when porting to other platforms. The advantage is great speed. The disadvantage is higher maintenance costs.

From a cost of development point of view, it is best to develop applications that target managed environments such as the Java Runtime Environment and the .NET framework[2].

For the cost of a slight performance hit, you get more robust code, faster development times, and a rich API library.

However, sometimes performance is more important to customers. As long as the customer knows the tradeoffs, then all is good.

> **Comment:** Microsoft's Universal Windows Platform (UWP) applications are compiled to native code for the platform they are installed on. You focus on the application and the system takes care of the rest.

> **Comment:** Some mainframe companies pay huge salaries to people who can code in assembly (Native language of all computers). Every picosecond of time is gold to them.

Machine Memory Management

Memory is a finite resource. There are several ways we can run out. This will negatively impact performance as the operating system tries to manage requests for more resources than it actually has.

Some causes for excess memory usage includes:

---

[2] The .NET framework is available for non-windows operating systems. See *www.mono-project.com* for details.

Memory leaks
Memory will be wasted if you forget to destroy objects when no longer needed.

This affects unmanaged applications the most. However, this affects managed applications as well.

Solution: Remove references to objects you no longer need.

 Warning: An object might have a reference to a large object that isn't needed. This is magnified if you are dealing with collections. Sometimes it is better to recreate objects as needed.

Requesting more memory than is available
The OS can make it appear that you have more memory than you do. If the OS runs out of memory, it will write your data to disk. It will then swap the memory as needed.

The problem is magnetic hard drives are many orders of magnitude slower than computer memory. As a result, your program will suffer a major performance hit.

Modern cloud servers and modern desktop and laptop machines use solid state drives (SSD). They give applications a huge performance boost. However they are expensive.

Solution: Employ a memory manager for your data collections.

Multi-Threading, Multi-Processing
Many problems can be broken down into sub-problems that multiple agents can operate on at the same time. By letting multiple agents operate in parallel, we can speed up the time we need to complete an operation.

Multi-threading is essential for UI responsiveness. The user gives a command and an agent, operating on a separate thread, responds to the request. This allows the user to do something else without having to wait for the previous command to complete. Once the agent completes the task, it invokes a callback function, which then processes the results. This will then update the UI with the results.

However multi-threading can cause problems when modifying share resources.

 Take Home: Use read-only collections of data in multi-threading application.

Take Home: Use shared data only for accumulators. Accumulators include filesProcessed, totalProcessTime, etc. However, accumulators should be implemented as managed objects that ensure data consistency.

Network

The network is by nature an unreliable resource. Performance depends on the network speed, as well as on how busy the web service is. There is always a danger that the connection may break.

The solution is using a proxy (See p. 94).

Avoiding using too much bandwidth gives two benefits:

- It speeds up transaction times. The fewer packages we send, and the less often we send them, the less time we need to spend waiting for a reply.
- It conserves resources. This is important for when bandwidth metering exists.

Scarce Resources

Scarce resources include database connections, working memory, and CPU cycles. For such situations, it is essential to monitor usage using performance counters and analyze if changes are necessary.

3.2. Availability

God does not begin by asking our ability,
only our availability,
and if we prove our dependability,
He will increase our capability.
-- Neal A. Maxwell --

Availability is a matrix that is mostly only relevant when dealing with distributed systems. Can I use the web site or is it down for maintenance?

Many things affect availability. Fortunately there are solutions for handling these issues, at **H.A.N.D.**

3.2.1. Hardware

- Hardware, such as the hard drive, could fail.
- There could be a power failure.
- The data center could burn down.

3.2.2. Available Services

- The application could hang and need to be restarted.
- The application needs to be shut down for patching.
- The operating system needs to be restarted

3.2.3. Network issues

- A tree could fall down and knock down a communications line.
- The network could be pushed to capacity

These are thing we have little or no control over.

The only solutions are:
- Schedule maintenance for times when demand is lowest.
- Use distributed computing

3.2.4. Distributed Platforms

One way to avoid downtime is by using server farms, where multiple servers run the same application. Services remain available even if individual servers go down.

Maintenance can then be performed on a rolling basis.

Of course the building could burn down, which means that if you want good availability, you need to have your services hosted in multiple locations.

That's a lot of work. Fortunately, there is a solution.

Infrastructure as a Service (IAAS)

There are multiple vendors that offer IAAS, such as Amazon Web Services, Cisco Systems, and Rackspace.

These companies offer to host your solutions on their infrastructure. They rent out their infrastructure and you install whatever operating system and software you need on blank virtual machines (VMs).

You define how you want your components installed and they run the script when needed.

Platform as a Service (PAAS)

Companies like Microsoft go a step beyond that and include the operating system. They update the servers as needed. The server is of course

Windows Server. The only drawback is they tend to be slow upgrading to the latest version of Windows Server.

As with IAAS, you define how you want your components installed and they run the script when needed.

 Warning: With IAAS and PAAS, the vender has the option of decommissioning the server without warning. When this happens, they create a new instance and then run your install script.

This means all persistent data will be lost without warning.

Take Home: When using an IAAS or a PAAS service, you are responsible for using a separate persistent storage solution for all your data needs.

3.3. **S**ecurity 1 (Managing Users)

Relying on the government to protect your privacy
is like asking a peeping tom to install your window blinds.
-- John Perry Barlow --

We create software to serve the needs of our customers.

Unfortunately the world is filled with malicious people who enjoy messing with our users.

Even without bad guys, having user groups simplifies the use of a product. People only see what they need to see.

Before we can begin creating applications, we need to know who our users are and what their needs are. This will allow us to limit who can do what.

As a general rule we have single-user applications and multi-user applications. Web applications are the multi-user type.

For single-user applications, we just have to keep out the bad guys. See – *Security 2 (Keeping* Bad Guys (P. 39)

Security is serious meat and potato business – so have a **S.P.U.D.**

3.3.1. **S**ecurity Groups

To simplify managing users, we have security groups. Each group defines what a specific set of users is allowed to do.

Single Responsibility

Each security group must represent a single management role. This role is defined by how the users interact with the product.

Roles include:
- Support
    - Tier 1
    - Tier 2
    - Tier 3
- Customers
    - Free Subscription
    - Basic Subscription
    - Premium Subscription
- Management
    - System
    - Network
    - Website 1
    - Website 2

 Note: User groups need to be defined at a management level, since it defines who uses the product.

3.3.2. Permissions

The purpose of permissions is to define what a specific group of users has permission to see and modify.

All functionality can be grouped under various categories and sub-categories.

Permissions are just flags that tell the application whether to enable specific functionality.

Single Responsibility

Each permission represents one and only one action.

Examples include:
- Users
    - View Users
    - Add user
    - Remove User
    - Modify User
- Content Developer

Yes, this means we can have potentially dozens or hundreds of permissions. That's okay. This will simplify the application in the long run.

 Note: Permissions need to be defined at a management level, since it defines who uses what features of a product.

3.3.3. Users

A user is anyone who has permission to use the system.

Users must **never** have permissions assigned directly to them. Doing so will make the solution harder to maintain, and potentially cause a security hole.

A user can make any changes that the permissions on the security group they belong to allow. This includes creating and modifying content. However, everyone else in the group has the permission to modify this content. The exception is user data.

Single Group
A user may only be a member of one security group. If a user can't fit in an existing group, then a new group must be defined.

This will simplify user management and product maintenance.

3.3.4. Data

All users have personal data that include their system preferences, profile and user personal content. But can it be held in a **C.U.P.**

1. Configuration Data
Configuration data allows the user to personalize how the system appears to them. Only the user needs access to this and this data will exist only as long as the user exists.

2. User Content
User content is any content the user produces that's not part of a shared media.

Shared content:
- WIKI
- Source Control contents

User Content:
- Personal internal blog

- Personal network storage

Personal content belongs to the personal. A decision needs to be made as to what to do when a user leaves the system. Will it be assigned to someone else or will it be deleted?

3. Profile

A user's profile is used to help others get to know the user. The system needs to allow a user to decide which parts of their profile to show the colleagues, the company and the world.

Of course, some parts of a user's profile can't be modified by the user. However, there needs to be a business reason to prevent a user from modifying their own data.

 Note: User data is data specific to the user and not the group.

3.4. – Security 2 (Keeping Bad Guys Out)

Just as drivers who share the road must also share responsibility for safety,
we all now share the same global network,
and thus must regard computer security as a necessary social responsibility.
To me, anyone unwilling to take simple security precautions is a major,
active part of the problem.
-- Fred Langa --

We will bankrupt ourselves
in the vain search for absolute security.
-- Dwight David Eisenhower --

Warning: Criminal hackers are everywhere trying to steal your data and your money. If you don't do anything your money will be gone. Your secrets will be used against you. Your intellectual property will make others rich. You could be accused of crimes you didn't commit. The world will come to an end.

Security is something that has to be built into a product from the ground up. Security is like insurance. It seems like a waste of time and money, until something bad happens. However, when that bad something happens, we are thankful for that insurance and the lock on that door.

It's time to **S.T.R.I.D.E.** through the dimensions of security.

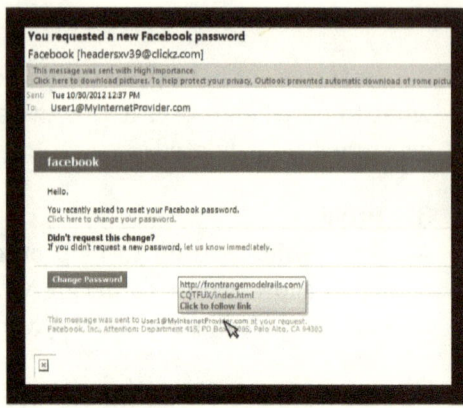

Web request to reset a user's password

3.4.1. Spoofing

Spoofing is the act of pretending to be someone the victim trusts. The bad guys then use that trust to take you to the cleaners.

Recently I've been getting emails from a certain wireless career claiming I owe them money.

I knew it was a phishing email because:
1. I don't do business with them.
2. I only use that email address for one specific social group I belong to.

So how does that work? Messages move from one server to another server until they reach your mailbox. The servers have the built-in ability to track where the messages came from. This feature is usually turned off. As a result, hackers can fake the originating address.

Only an act of congress can change that.

Business Verification

To counter spoofing, some companies, such as the Bank of America, requires you to select an authentication image when logging onto their web site. The web site then displays the image on its login screen, with a warning not to log on if you don't see the image.

Business Signature
See this image,
and know,
you have entered
the correct web site

Warning: I recently began getting bounced emails from my email account. The account was sending spam emails for some unknown reason. I changed my password and the problem disappeared.

Somehow my password was compromised and bad guys were using it to send spam.

Take home: Regularly update your passwords.

3.4.2. Tampering

Another word for vandalism, tampering is when someone alters or destroys data for malicious purposes.

There are two ways to guard against tampering: Backing up data and limiting access.

Limiting write access to only those who need access greatly reduces the change that data can be tampered with.

This is solved using security groups that specify what permissions individuals have.

Backing up data on a regular basis allows us to repair any damage caused by tampering.

SQL Code Injection

There are two ways to retrieve data from a database: Ad-hoc queries and stored procedures.

With Ad-hoc queries, the user creates a string and submits it to the database. If the string is generated at the client side, then the client can do anything the security settings of the database allows. In the worst case scenario, the person could wipe out the database or steal corporate secrets.

Stored procedures are the only recommended way of retrieving data when dealing with user input. With stored procedures, the admin decides what the user can do and see.

Stored procedures have a second advantage over ad-hoc queries – speed. A stored procedure can be ten times faster than an ad-hoc query.

 Warning: Some internet solutions just strip away certain characters from user inputs as a safety precaution.

While this works, it negatively impacts user experience.

Website Defacement

Using the above code injection, a hacker can damage a web site. However, you usually need an account and password to do other kinds of damage. But not always — Some sites allow un-authenticated users to add comments, etc. through their site.

3.4.3. Repudiation

Repudiation simply means being able to do bad things secretly. No one knows who you are.

The countermeasure to repudiation is Authentication. Authentication is the act of verifying a person's identity. There are multiple ways to authenticate a user. Here are the most common.

Passwords

This is done through user name and password. Like all things in life, passwords have strengths and weaknesses.

The strength lies in the fact that it is easy to implement and is a standard way of authenticating users.

It however has several weaknesses. These include:
a) Passwords can be stolen
b) Passwords can be hacked
c) Passwords can be mistyped or forgotten

For more details, see Password Security in the appendix (p. 165)

Smart Cards

Smart cards are popular in government and business. The most common use of smart cards is access to buildings. The employee presses their card next to a reader. The system then accepts or rejects access based on its own criteria.

Some businesses use smart cards to allow employees access to its corporate network. An example is Microsoft. The employees place their card in the reader and with the use of special software and a user password, they are granted access to the corporate network.

The strength of using a smart card is that it can be hard to hack, when properly implemented.

The weakness is that cards can be lost or stolen.

> "A PC with a Smart Card reader communicates through this gold contact to the microprocessor. If the authentication data stored on the Card's programmable ROM (Read Only Memory), such as the username and password, matches that of the PC, the user is granted access. Beyond merely allowing access to a PC, a Smart Card securely encrypts and decrypts individual files on the PC.

> "The drawback to a Smart Card is that you have to remember to bring it with you at all times. If you head out on a business trip with your laptop, but leave your Smart Card on your desk, you won't be able to use your laptop until you get back to the office. Smart Cards can be stolen along with your laptop. According to Tom Aebli, director of PC segment marketing for fingerprint sensor manufacturer AuthenTec, Smart Cards work well, but you don't know who's using them. 'I can give my Smart Card to my buddy, go away on vacation, and ask him to check my e-mail while I'm gone. That's a concern for IT managers.' "[3]

Radio Frequency IDs (RFID)

Radio frequency identifiers use induction to communicate over short distances. The advantage of near field devices is that batteries aren't needed.

Smart cards are an example of RFID devices. You place the card next to the reader and the reader reads the card.

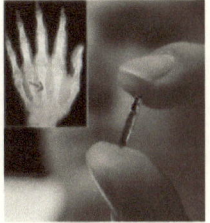

RFID devices can also be placed in other items, such as a ring or watch. It is a convenient solution, since people can wear the ring at all times.

MichaelJournal.org/RFID.htm

An RFID device can be contained in a glass tube that can be embedded under the skin. This is often used to track livestock and pets.

However some people have chosen to embed them in their hands as a replacement for smart cards. The advantage is that you can't lose them. The disadvantage is making sure a reader doesn't read it when you don't want it to.

 Note: It is forbidden for companies and governments to use embedded RFIDs for religious, moral, ethical, and in some instances, legal reasons.

Biometrics

A common way for authentication is fingerprint recognition. Some governments also use eye scanners.

- http://archive.laptopmag.com/Features/Real-Notebook-Security.htm?Page=2

Another way for authentication is using the vein pattern in your hand. The vein pattern in a person's hand is unique to that person, since the exact pattern isn't fully determined by genetics.

An infrared scanner scans your hand and uses that scan as a means of authentication.

Open ID
In the words of Wikipedia[4]:

> OpenID is an open standard that describes how users can be authenticated in a decentralized manner, eliminating the need for services to provide their own ad hoc systems and allowing users to consolidate their digital identities...

> The OpenID protocol does not rely on a central authority to authenticate a user's identity. Moreover, neither services nor the OpenID standard may mandate a specific means by which to authenticate users, allowing for approaches ranging from the common (such as passwords) to the novel (such as smart cards or biometrics).

Open ID is used everywhere on the internet for authentication. Examples include:

- Aol.com
- Blogger.com
- Flickr.com
- Google.com
- Hyves.com
- LiveJournal.com
- MySpace.com
- Orange.fr
- WordPress.com
- Yahoo.com

For more information, see: http://openid.net/

The drawback of such a scheme is that once an account is compromised, all accounts are also compromised.

3.4.4. Information Disclosure
Both people and organizations have information they need to keep private. If permissions aren't properly set, then outsiders may gain access to restricted information.

To guard against this, we need to:
1. Identifying the business and customer information within the system
2. Define the rules for who can view and change the information.

---

[4] http://en.wikipedia.org/wiki/List_of_OpenID_providers

3.4.5. Denial of Services

In a denial of service attack, a hacker floods the service with countless bogus requests.

The main solution here is to respond only to legitimate users and keep a transaction log. We then analyze the service requests to see any suspicious behavior.

A good way to handle denial of service attacks is to have a guardian service that verifies that requests are legitimate before passing requests on.

 Comment: A Distributed Denial of Service (DDOS) attach is just a denial of service attack (DOS), launched from multiple IP addresses.

3.4.6. Elevation of Privileges

In an Elevation of Privileges attack, the system allows a person with insufficient authority to view data and make changes that in theory they aren't allowed to make.

There is only one way to deal with security (See p. 36).

You need to ensure:
1. Security Groups exist for all roles
2. Permissions are defined for all operations
3. Users are defined
4. Data is managed

 Warning: Every action performed on your system needs to be associated with a user right. This will allow granular access control for all your functionality.

 Comment: This is your intellectual property. The inability to control it can cause your business to go bankrupt.

The end of Vulnerabilities? – Hardly.
Only a non-functioning computer is safe from being hacked.

3.5. Scalability

There's no limit possible
to the expansion of each one of us.
-- Charles M. Schwab --

We live in a world where we can have thousands of people visiting our web site at the same time. With bigger sites such as twitter and YouTube, we have millions of simultaneous requests.

Previously we talked about performance. With performance, we speed up the processing of requests by making the code more efficient. We then scale up the hardware to enable greater resources.

However, there is only so much you can do to increase the performance of a service.

Scaling up is no longer a business strategy. Manufacturers don't know how to increase speeds without using crazy amounts of electricity and cooling.

The new solution now is to build out. Building-out means dividing the workload between multiple machines with multiple cores in a server farm.

Note: Moore's Law says that the number of transistors on an integrated chip will double every 18 months.

This does **not** mean that speed will increase every 18 months. The link was just assumed. Smaller transistors used less power, and so could be operated faster. We

can't build smaller, so the assumption no longer holds.

The real power of Moore's law is creating chips with ever increasing numbers of cores. The big winners are applications that benefit from parallel processing, such as: Graphical Applications, Fully Immersive VR Environments, Big Data, Server Farms, and much more.

Solutions of the future must use **M**ulti-server clusters, **A**pplication Pooling, and **D**ata Warehouses to enable solutions that can address the needs of the world. If the scalability issue can't be addressed, then people might get **M.A.D.**

 Comment: Scaling-up just means buying more expensive equipment. Scaling out is using more agents to do more work.

3.5.1. **M**ulti-Server Networks

The traditional way to handle higher demand is to buy additional servers and infrastructure. Your maintenance crew then installs the operating system and necessary software.

The advantage is that you have full control over the entire system.

The disadvantages are:
- You need to hire people to maintain the system
- You need to over-provision to deal with peak demands
- Your investment gets underutilized most of the time

With cloud services:
- There is no need for dedicated maintenance personal
- You only pay for what you use
- You can quickly scale up at times of peak demand.

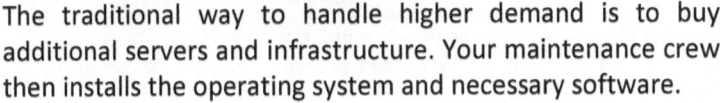 **Comment:** A trend in cloud computing is for service providers to buy servers containing large numbers of cores. They then use virtualization to create standardized virtual machines. This allows for greater control of resource allocation.

Combine this with solid state drives (SSD) and we have a solution that offers both speed and flexibility.

3.5.2. **A**pplication Pooling
With application pooling, we have multiple instances of the application running at the same time. These instances can be running on the same machine or different machines.

When a request comes, it is routed to the next machine that is free. A gatekeeper routes requests as needed.

The most important requirement for the application is that it be stateless. All state resides in separate databases. This is because there is no telling what server will respond to a request. Also, servers can be decommissioned without warning.

3.5.3. Data Warehouses

Data warehouses store your application state, auditing information, customer and business data, among other things.

This separate data storage is essential to prevent data loss. The reason is because you can't rely on a specific instance of a server being alive indefinitely.

 Take Home: People are embracing cloud computing, Infrastructure as a Service (Ex: Amazon Web Services), and Platform as a Service (Ex: Microsoft Azure) for their web and big data needs.

That's not surprising. They don't want to go **M.A.D.** dealing with hardware failures, too much\too little hardware, operating systems updates, necessary restarts, database failures, and the need for dedicated personal, to name a few issues.

3.6. Maintainability

I couldn't repair your brakes,
so I made your horn louder
-- Auto mechanic saying --

Maintainability simply refers to the ability for new developers to come in and make necessary changes with a minimum amount of disturbance to the system.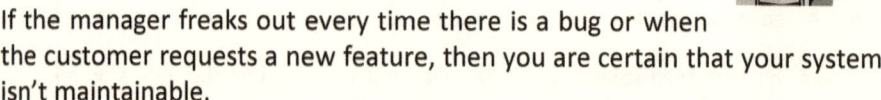

If the manager freaks out every time there is a bug or when the customer requests a new feature, then you are certain that your system isn't maintainable.

Only a **D.O.L.T.** forgets about application maintenance.

3.6.1. Documentation

Documentation is essential when it comes to maintainability. The ideal way to create documentation is to write it for someone who has never seen your project, then let them review your project.

Internal WIKIs are excellent tools for managing documentation.

3.6.2. Object-Oriented Programming

Object oriented programming is all about maintainability. It makes real-world problems more manageable than otherwise.

However, it's just a tool, just like any other tool. You need to understand how and when to use it, or it is useless. The object-oriented paradigm will be discussed in more detail on page 69.

3.6.3. Logging

Logging allows you to keep track of the flow of an application. With this you can trace when the application doesn't behave as expected.

 Warning: Log and Trace files potentially contain security information. Procedures must be in place to handle them.

 Comment: Microsoft has an issue tracing system called Dr. Watson. It is used to report software crashes to Microsoft.

I wonder if there is a Sherlock Homes somewhere looking over the files, trying to catch evil bugs.

3.6.4. Test Automation

A major key in creating manageable code is proper verification code. This gives us a certain degree of confidence that we didn't break functionality when we make changes. Automation will be covered later (See p. 146)

3.7. Accessibility

Basically, our goal is to organize the world's information
and to make it universally accessible and useful.
-- Larry Page --

Accessibility is a dimension that deals with how people with limited mobility, vision and hearing interact with your application.

When dealing with this issue, we need to know who our customers are and what their needs are. Failing this will cost us in lost opportunity. Let's be **H.I.P.** to the needs of our customers.

3.7.1. Hearing

For hearing impaired people, we need to make sure any audible signal has a visual counterpart too. All verbal instructions need to be available in written form.

3.7.2. Impaired Vision

For visually impaired people, hearing can act as their eyes. In this case, we need to make sure that the relevant messages can easily be translated using existing speech tools.

Some tools are designed to help us control our computer using speech. Our software will handle this if we use standard controls, and program to publish standards.

However, we need to test to make sure our targeted audience is being properly served.

When dealing with vision, there are three things to keep in mind.

Color-Blind

Some people are color-blind. That means that you shouldn't rely on color to convey information, unless there is a way for the person to customize the colors.

There are accessibility programs that mimic color blindness by adjusting the colors of your screen. This allows you to see what color-blind people see.

Flashing lights

Applications sometimes use flashing images to catch people's attention. For some, this is just annoying (Yes I saw that stupid message. Quit bugging me). For others, it can cause a seizure. Enough said.

Feature Size

Some people will find it difficult to read text that is too small. The same is true for icons and other visual features such as images.

Web browsers have a feature that allows the user to scale the displayed pages. This is achieved by holding the <CTRL> key and turning the mouse wheel. You should consider this instead of forcing the user to rely on magnifying glass programs that magnify parts of the screen.

Tools that adjust the resolution or that blur the image are great for showing us how our customer views our product.

Custom Controls

Custom controls are different from standard controls in an important way. Standard controls are designed to work with standard accessibility tools.

How do custom-controls work? You start with a blank control and you draw whatever you want. As a result, accessibility tools will not understand how to show necessary information to users.

 Take Home: When creating custom controls, always follow the standard guidelines when it comes to accessibility.

Warning: Don't use custom controls for security. It will only limit customer access. It will not block crackers from cracking your system. Never underestimate a dedicated hacker or cracker or the power of AI.

3.7.3. Phones

If your application is hosted on a web browser, there is a good chance that the user might access it with a cell phone.

There are several challenges associated with cell phones.

Screen Size

Cell phones have tiny screens.

The solution phone browsers use is:

- Render the web pages super-tiny. This is annoying for even for people with standard vision.
- Use scroll bars. Having to constantly scroll is also annoying.

The solution is to use dynamic controls that rearrange content based on screen form factor.

Control Size
We are using fingers to control the UI. That means finger size will affect how well we can use the web site.

Mouse-Over Actions
Cell phones don't have mouse pointers. That means that actions based on mouse movements won't work.

The only control option we can rely on is the finger touch.

An example of a control that can cause problems is the drop-down list. The drop-down list appears when you mouse-over the control. This is impossible for cell-phone users.

 Comment: In the past, most people had low resolution monitors. Applications were designed with that in mind. Now, applications don't worry about low resolution screens. With cell phones, we have come full-circle.

Take Home: HTML5 libraries exist to dynamically reconfigure the page for best viewing on various screen form factors. However, you must make sure they look good on the supported devices.

3.8. Deployability 1 (App Types)
See, unlike most hackers,
I get little joy out of figuring out how to install
the latest toy.
-- Jamie Zawinski --

An application is useless unless its functionality is made available to the customer. Deployability deals with how you package an application and deploy it. It also deals with updating an application in a way that is convenient for the customer.

There are two main types of applications: Web applications and platform applications.

3.8.1. Web Applications

Web applications are relatively simple to deploy. As an example, if you are working with an ASP.NET project and Visual Studio, then all you need is permissions on the server and you can easily deploy.

The tricky part is dealing with persistent storage. However that is one of many services your system will use.

3.8.2. Platform Applications

Platform applications are more challenging. The number of platforms is large and growing.

Platforms include:
- Mobile
- Desktop/Laptop
- Servers
- Smart devices
- Cars
- Etc.

There are multiple ways to deploy your application:
- Through your own web site
- Through popular download sites
- Through the app store of the platform you support
- Using physical media such DVDs

There are:
- Single user applications
- Server-based applications
- Multi-server/Cloud solutions

3.8.3. Single-User Applications

Single-user applications are typically installed on a single machine. This could be on a home computer, laptop, tablet, phone, etc. The installation completes and the application is good to go.

When dealing with deployment, we have four issues to deal with:
- Install

- Upgrade (Service Packs/Bug fixes)
- Repair
- Uninstall

3.8.4. <u>Multi-Server Applications</u>

Multi-server solutions can be a pain to install. For each server, the admin needs to install and then configure the service. They then have to configure the entire solution so the servers communicate correctly with each other.

The solution here is to use Virtual Machine snapshots and Docker for third party components. This will streamline the installation process.

3.8.5. <u>Cloud-Based Applications</u>

With Infrastructure as a Service (IAAS), solutions can be as complicated as multi-server solutions if all the service provider provides is the infrastructure.

You get a blank VM. You then specify the operating system, support software and configurations. Finally you install the application.

With System as a Service (SAAS), the operating system is given. The provider manages OS patches. You just need to worry about your application and support software.

For both IAAS and SAAS, you supply an installer. The infrastructure runs the installer as needed. This is either on demand or when the service decides to reallocate resources.

To update the service, simply upload a new installer. Then make a request for the system to run your installer. Or you can just wait until they system automatically updates.

3.9. – <u>Deployability 2 (Platform Installation)</u>

For many people my software is something
that you install and forget.
I like to keep it that way.
-- Wietse Venema --

Installers allow customers to control how an application is managed on their system. This includes: Install, Update, Repair, and Uninstall.

3.9.1. <u>Installation</u>

Ideally, we ship an installer to the customer. The user runs it and after entering some necessary information, the application is ready to use. Here are a few things to keep in mind when creating an installer.

<u>Quick Start</u>

The installation wizard start-screen should appear the moment the user opens the installer. At this point, you can notify the user that the installer is collecting data for the installation of the product.

A good example of where this fails is when you install Microsoft SQL Server. You click on the installer and nothing happens. You then wait, wondering if the installation started. You restart the installer and then you find multiple installer windows.

<u>Fail Early</u>

All applications have prerequisites for installation. These include:
- System requirements
  - User permissions
  - Preinstalled software
  - Hard drive space
  - Etc.
- Product keys
- Product activation. It is best to mention it at this point, before the product installation begins.

For the convenience of the customer, if the installation is going to fail, then it is best to fail as early as possible.

<u>Upfront Initialization</u>

This is similar to the fail early philosophy. For the convenience of the customer, we should collect all necessary data as soon as possible. Many times, this data will determine if an installation will fail.

By finishing all necessary user steps in the beginning, the user is free to ignore the installer until it has finished.

<u>Blocked Installation</u>

Many times the installation of a product is blocked because the pre-requisites aren't fulfilled. At this point, the user has no choice but to abort the installation.

When this happens, you should allow the user to abort without asking to confirm that they want to abort. Such a question is just annoying, since they have no choice in the matter.

Installer: "Installation can't proceed. Do you want to abort?"
User: "Abort Installation"
Installer: "Are you sure you want to abort?" (Even though you can't do anything?)

Status Messages

Always keep the user informed of the status of the installation. It's annoying for the user to just wait, wondering if the installation is proceeding properly or if it is blocked.

However don't use blocking message boxes for status messages. People need the option of leaving and coming back as needed. As mentioned, collecting data first makes for a better user experience.

Unnecessary Steps – (Keep it Simple) KISS

Many times the installer forces the user to walk through page after page of information that has nothing to do with the installation of the product.

If you insist on showcasing how incredible your product is, wait for after all configuration steps are complete, and then show what you want to show.

However, don't force the user to click on buttons, just to make them see your show. This is unnecessary, since they already bought your product. Also, if your installer has video, allow the user to mute the sound. Better yet, link to a YouTube video.

Sometimes installers have pop-ups that say the installation is complete, regardless of whether the installation completes or fails. This is just annoying since it adds no new information.

This is especially annoying when the user aborts an installation. First the user is asked to confirm the abort, and then a popup appears saying the installer has finished (Which it hadn't. It was aborted).

Reuse Windows

A typical installer requires the user to navigate through multiple pages before installation is complete.

Some installers use separate windows for each page used. As a result, when you press the next button, the window disappears and a new one appears.

I find this behavior annoying. This is especially true when I want to move the installer screen to a location other than its default location at the center of the screen.

I move the window to the side so I can do other work while the installer runs. I press the 'next' button and the window disappears, only to reappear where I don't want it to appear. I then have to reposition the window so I can do other work.

 Take Home: A typical user runs multiple applications at the same time. They might even run several installers at the same time. Smooth installations make for happy customers.

Side-By-Side-Installation
Software changes from version to version. That is the nature of software. Unfortunately the result of this is that documents produced by one version of a product may not be compatible with another version.

In addition, the user interface can change greatly between versions.

As a result, some people like having multiple versions of the software running on the same machine.

Having products installed in version dependent installations is a simple way of dealing with multiple versions of the same product.

In the past, the cost of hard drive space was high. As a result, it was important to share resources.

Now, it can be more cost effective to just duplicate all resources. This way one installation will not potentially break another installation.

 Comment: Consider allowing users to select install directory.

When given the option I always switch to D:\Program Files\<Program X>. The reason is that my C: Drive is on an SSD with only 250MB of space.

3.9.2. <u>Update (Bug-Fixes, Service/Feature Packs)</u>
All programs have bugs. That's the reality of the computer world. The question is how do you deal with it?

Many companies publish updates to their products that fix bugs and add functionality. The simplest way is to just replace the assembly with the new

one and then update the product configuration to handle the new assembly.

When publishing updates, you have two options: Do it automatically, or make the user perform some actions.

Adobe displays a message whenever an update is ready. The user then has to agree to the contract and step through the installer every time there is an update.

I personally find this annoying. I already agreed to the contract when I first installed the product. Why do I have to do this every week or so? I understand if the terms of agreement changes. However, how often would that change?

Obviously the best solution is to leave this as a user configuration. The user can then decide if they want updates. If updates are requested, the user should be allowed to have either silent updates or manual updates.

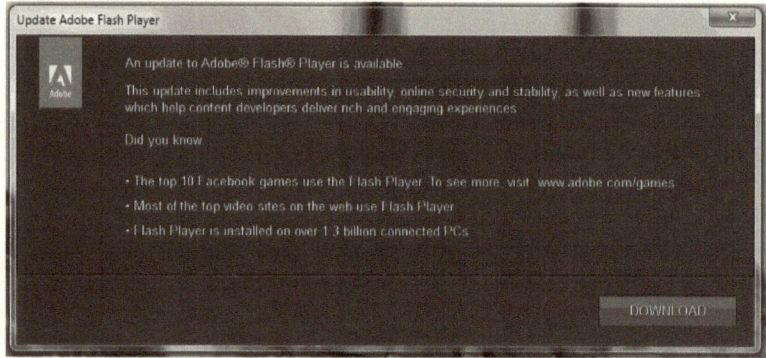

Dialog informing user an update is available.

3.9.3. Repair Installation

Applications stop functioning for various reasons. This could include corrupted or missing files or entries in the registry.

As a result, a useful piece of functionality to have for you installer is to be able to repair an installation. Of course, this repair process needs to take into account user configurations and published updates.

3.9.4. Clean Uninstall

When uninstalling, it is best to remove all traces of a product. Configuration files can be stored at a customer-specified location.

User data is different. The installer must ask the user what to do with the data. Just leaving the data is not an option.

3.10. Extensibility

Adding functionality is not just a matter of adding code.
-- Wietse Venema --

How well is your application designed to handle the addition of new features?

Extensibility is related to maintainability. Just like maintainability, extensibility relies on the use of proper development practices to work.

3.10.1. Plug-in model
Extensibility can be achieved using a plug-in model. You have one central executable and multiple DLLs that contain your functionality. Each DLL would specify the commands that the program needs to expose and the context for the commands.

Adding new functionality would be as easy (in theory) as adding a new DLL to the DLL folder. The app displays the new content based on information found in the DLL's metadata.

3.11. Responsiveness

The point is that you want
to have a system that is responsive.
-- Bill Joy --

Responsiveness isn't about speed, but the perception of speed. As such, it is related to usability.

Users are willing to wait if they know that something is happening.

However, when the application takes too long responding to our requests, it is like being stuck in a traffic jam. We just waste gas, as the application **B.U.R.P.**s along.

3.11.1. Busy Indicators

I don't know about you, but it drives me crazy when an application just sits there and does nothing. Activity indicators show us that the application is doing necessary work and not just stuck.

Busy and activity indicators are good aspect of UI design, and show the user that we aren't wasting their time.

3.11.2. UI Responsiveness

Under no circumstance should the UI become unresponsive.

An example of an unresponsive application is Microsoft Word 2007. Word 2007 is designed to auto-save every ten minutes. When this happens, the UI becomes unresponsive for 5-10 seconds. In the meantime you have lost your train of thought, to say nothing about the fact that the wait is annoying.

 Take Home: The UI thread should never be used for anything other than managing the UI.

3.11.3. Resource Management

Resource management is essential for larger programs. Increasing thread count and caching data can speed up response times – up to a point. However, using more resources than are available will greatly slow down your program.

Memory leaks can kill performance.

When dealing with resources, you need to always know what is available, so you can plan appropriately.

 Take Home: Using a resource Manager is essential when dealing with a resource hungry app.

3.11.4. Program Startup

Program startup can be a time consuming process. We need to handle this situation properly or we will have dissatisfied customers.

Startup Screen
The purpose of a startup screen is to tell the user that the application is being initialized. The first thing that happens on application startup is you show the startup screen. Next you start the needed modules required by the app.

The ideal start screen shows startup progress, allowing for the perception of greater responsiveness.

As a (bad) example of this, I refer you to Microsoft Outlook 2007. At startup, the UI is displayed. However, it is not responsive to any user input. You can't even maximize or minimize the window.

One answer to this is the Command design pattern, discussed on page 157.

 Take Home: Never use the UI thread to perform business logic. This is especially true when connecting to remote services.

Lazy Initialization
Lazy Initialization is another way to improve the responsiveness of an app. You only load components when you need them.

The program starts fast. However, there is a time delay every time you request new functionality. The more modules there are the more annoying the process is.

When dealing with these programs, I sometimes wish they would just initialize everything at the start. That way I could take a coffee break, then come back and become productive.

Hybrid Initialization
Clearly neither all-at-once initialization nor lazy-initialization is adequate.

The fix is to use hybrid initialization:
1. Start the app fast.
2. Start modules as needed, as defined in lazy initialization
3. Use a resource manager to:
    3.1. Create an ordered list of modules that the user required.
    3.2. In the background, start modules one by one based on usage.
    3.3. Track available resources to prevent loading too many modules
    3.4. Unload modules when resources become limited

 Take Home: For maximum responsiveness, start the app fast. Then, in the background, start the most commonly used modules one by one.

3.12. Usability

We wanted to make our product
easy for our customers to use,
so we removed all functionality.

How many of us have worked with software that's
overflowing with incredible features, but is next to impossible
to use?

Many times the programmer worries about technical
correctness, without realizing that the user doesn't care about that.

The ideal program is one that even our **D.A.D.S.** can use.

3.12.1. Design Elements

The way the UI is designed greatly affects the usability of an application.

Minimalistic Design

Microsoft and other companies are driving a minimalistic design philosophy
where as many so-called unnecessary design features are removed, leaving
only bare functionality.

Windows App Store applications are a perfect example of minimalistic
design.

Keep in mind that having just one text box that does everything is doesn't
necessarily make it easy for the user to use.

A perfect example is the GPS application on Google's Android OS 4. At first
the application was relatively simple to use. Then they replaced everything
with two text boxes, one for the starting point[5] and one for the end point.
Under that box is a list of saved locations.

That lame UI made the application cumbersome to use.

The Familiar

When creating a user interface, strive for a look and feel that the user is
familiar with. This will simplify how the user handles a situation.

---

[5] Why do you need to specify a starting point for a GPS *phone* application? Is it
possible to start a trip at a location other than where you are?

When adding new functionality, make sure the feature adheres to the existing design pattern.

Skinning

Skinning is a technique where you give the user the ability to manage the look and feel of an application.

This will allow the developer to create new UI layouts, without fear of upsetting users. Users can then either use an existing skin or use the new one.

Design Patterns such as {Model-View-Controller} and {Model-View-Viewmodel} were designed to simplify the creation and management of custom views of your application.

Keyboard shortcuts

Keyboard shortcuts allow users to control an app without relying solely on the mouse. This is good, since constantly switching back and forth between mouse and keyboard can be a drag.

Also, excessive use of the mouse can hurt the wrist.

There are several keyboard shortcuts that are used across many applications. This allows people to leverage existing knowledge to work with your application. See p. 161 Keyboard Shortcuts for a list of common shortcuts.

 Take Home: Always use common control patterns. This will simplify user interaction, and make the application more enjoyable to use.

3.12.2. Annoying Popups

Developers use pop-ups for two separate reasons: To force a user to take an action and to inform a user about an action.

When dealing with blocking issues, popups are probably unavoidable. However, do you really need a popup to inform the user a task is complete?

For information messages, a non-modal window (that disappears on loss of focus) would work, or a message at the bottom of the screen, depending on the type of message. Ideally this type of popup is controllable by the user.

OpenOffice.org Example

Apache OpenOffice.org Writer, Version 4 has a good example of the annoying use of a popup.

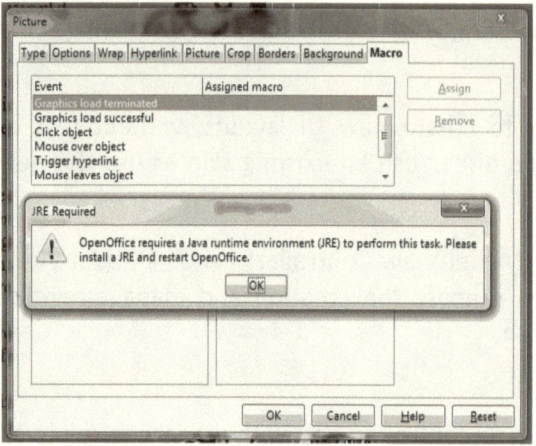

Issues seen here:

1. This message doesn't tell take into account that some customers aren't Java experts. This is a product designed for the general public, so this message would be meaningless.
2. This message appears every time you click on the 'Macro' tab.
3. This message showed up a dozen times.
4. Once you get through the popups and close the dialog, the next time you open the dialog it will happen again, since it was previously on 'Macro'.

For a situation like this, where action is needed to enable functionality, it is best to replace the tab contents with a message detailing the issue. Also include a help link.

3.12.3. Documentation

Documentation is essential for functionality the user is not familiar with. Try to be as concise as possible, when supplying help for specific tasks.

Microsoft Office used to have a contextual help button. You click on the button and then on the item you needed information on. The help system then gave you targeted information on what you selected.

Now all you get is general data, which you then need to search through.

<u>Tutorials</u>

When creating a tutorial, only instruct on one topic.

As an example, I saw a tutorial on codeproject.com[6]. In it the author gave a tutorial on how to create a WCF service. This author combined two tutorials into one: How to create a WCF service and how to use visual studios.

If the user wants to learn about using Visual Studio, the tutorial was perfect. However, for those who know how to use Visual Studio, the tutorial was cumbersome to read. It cluttered the tutorial with unnecessary information.

3.12.4. SDKs

Making an SDK for your application available might be a good idea. This will allow third parties to create cool skins for your application. This would increase the usability of your solution and potentially increase sales.

3.13. Globalization

Globalization is a fact of economic life.
-- Carlos Salinas de Gortari --

Markets for our products exist throughout the world. That's a lot of people willing to spend money on what we have to offer.

There's only one problem. They don't speak English. The solution is Globalization.

 Comment: Each culture has its own way of looking at the world and doing things. Expressions and images that are acceptable in one culture may be offensive in another culture.

The ideal solution is to get feedback from the targeted users.

Comment: Globalization is the process of enabling an application to support multiple regions.

Localization is the process where support for a specific region is added.

Care must be put into localization, since a bad job can put a few **L.I.D.S.** on our sales opportunities.

---

[6] http://www.codeproject.com/Articles/97204/Implementing-a-Basic-Hello-World-WCF-Service

3.13.1. Local Units

The format for time and money changes from region to region. The same is true for distance, weight, speed, etc.

When dealing with measurement, the important thing is to keep to one system and then translate when displaying information to the user.

3.13.2. Information Text

Text is an essential form of communication. However, it can cause confusion if not translated correctly.

Worse, some innocent sounding words or phrases can be offensive when improperly translated.

3.13.3. Daylight Savings Time

In many regions of the world, clocks are moved forward sometime in spring and moved back sometime in fall. As a result the timestamps of database transactions can be compromised.

It is best to record all time in universal time, and then convert the timestamp to suit your local needs.

3.13.4. Sound and Video

Images are powerful tools for conveying information. However, care should be taken when displaying images to customers in other regions of the world.

Many times an image that seems harmless to us can be offensive to others. In this instance there is no choice but to consult an expert that is familiar with the region.

This also applies to video. The subject of some videos could be offensive to some people.

 Comment: Mixing measurements can cause disasters that are out of this world. On September 23, 1999, NASA's Mars Climate Orbiter was lost because of it. See: http://mars.jpl.nasa.gov/msp98/orbiter/

3.14. Sex Appeal

If you can't make it good,
at least make it look good.
-- Bill Gates --

"Design has been viewed as being aesthetic. Design equals How Something Looks. You see this attitude to design in every part of society—clothing design to interior design, less so in product design, and yes, in web design.... I think design covers so much more than the aesthetic. Design is fundamentally more. Design is usability. It is Information Architecture. It is Accessibility. This is all design." — Mark Boulton

If a product doesn't scream out "Buy Me", then we need someone to pretty it up for us. It is a fact of life that looks are sometimes more important to a customer than substance. Having both is definitely a winning combination.

It slices. It dices.
It even does Julien fries.

 Comment: Customer satisfaction is more important than cool bells and whistles, which only annoy the customer.

On the other hand, properly designed bells and whistles can enhance the user experience.

4. THE MAP IS NOT THE WORLD

If I had a hammer,
I'd hammer in the morning
I'd hammer in the evening,
All over this land
-- Peter, Paul & Mary --

Software engineering is just a human endeavor, intended to help us solve problems in the real world. Different aspects of software require different approaches to solving. No one approach is greater or lesser than another. It depends on what you need.

The above Pass-Made-Rugs aspects of software are a perfect example of this.

As an example, some C/C++ software developers shun managed environments (ex. Java Runtime and the .NET framework) because C/C++ is faster, forgetting that there is more to programming than just raw speed.

Another holy war is between those who embrace dynamic languages such as JavaScript, and statically typed languages, such as C++.

Different stakeholders are concerned with different aspects of a business solution.

Businesses need solutions that bring in revenue:

Deployability	"How do we deliver the solution to our customer?"
Availability	"Time is money. If the service isn't available, then we aren't making money."
Globalization	"The world is a big place. How do we maximize our markets?"
Scalability	"Our business is rapidly growing. Can we keep up with demand? We don't want lost opportunity."
Security	"Our business data and user data is important and must be protected."

Marketing needs a way to sell the product:

Sex Appeal	"How do we make the look and feel of the program so awesome that it will crush the competition?"

The user wants a product that they can easily use:

Usability	"This is driving me crazy. How do I perform that operation? I did it before, but I don't remember now. This piece of (beep) isn't very intuitive.
Responsiveness	"This antique program is so slow, it makes glaciers look fast, and why does it freeze all the time?"
Accessibility	"I wish my grandmother could use this program, but the buttons are too small to see."

Strangely enough, the Object Oriented paradigm isn't necessary for these things.

Instead, the object oriented paradigm is intended for solution management.

Object Oriented Programming (OOP) allows us to deal with:
- Maintainability – Dealing with under-defined feature requirements (bugs)
- Extensibility – Adding new features

Notice I never mentioned performance? Performance is and should be the result of good design, taking advantage of the latest advances in software engineering. More will be said about performance, in terms of technology and algorithm choice.

4.1. Object Oriented Programming

It has been my experience that
what perplexes and frustrates many people
are the higher-level concepts
of object-oriented programming.
-- Dan Clark –

All books on software engineering have a section on Object Oriented
Programming, so here it is – but first some history.

4.1.1. Functional Decomposition

In the beginning of the computer era, when space and
time began, computers had less processing power
than our cell phones. At that time, a natural way
of solving problems came into being. It was
called functional decomposition.

> **Amiga 500**
> It was an incredible machine.
> Whenever it crashed,
> it would give out
> "Guru Meditation" numbers.

You take a problem and break it down into
smaller problems. Once you have broken down the problem into small
enough pieces, the solution, in theory becomes trivial.

I say in theory because some problems are circular. You end up where you
started. That is something many engineers aren't aware of, both then and
now.

Functional programming was perfect then, since most C/C++ programs
could be printed out in less than five sheets of paper – and often just one
sheet of paper.

Eventually, program size increased as the speed and memory capacity of
computers increased.

Flaws in this monolithic approach to software development appeared, since
it failed to take into consideration the hierarchy of functionality needed to
solve real-world needs.

Comment: Functional decomposition has a bad reputation, since people had
bad experiences with functional programming.

Functional decomposition is just the first step in achieving the goal of satisfying our
customer's needs.

4.1.2. <u>Code Reused – Birth of the Function</u>

Not surprisingly, people noticed that they were using the same code fragments throughout their program. Each code fragment always did the same logical task. If the code logic was wrong for the fragment, the programmer would have to make a revision in multiple locations in their program.

At first, to enable code reuse in simple C, people used *goto* statements (or at least I did) to place the common code in a convenient location. That was anything but convenient.

Language designers came to the rescue with program functions. This allowed common functionality to be isolated. With the use of input parameters, the functions became quite useful.

We still had a problem. Variables were shared throughout the program, making it hard to debug errors.

4.1.3. <u>(Pseudo) Object Oriented Programming</u>

Language designers came to the rescue again and added the concept of the *class* to C. This, along with a few other additions gave us C++.

> **Comment:** Contrary to public opinion, C++ isn't an object oriented programming language. It is a functional language, with program constructs that allow people the opportunity to work in an object-oriented way, if they put the effort into the process.

Classes allowed programmers to group common functionality with the data they operated on. It made the programs more robust to changes in design. It also simplified debugging, since only limited parts of a program could modify variables.

The concept of Object Oriented Programming came into being. The principle was simple. Identify the nouns and verbs in the problem you are working on and then wrap the functionality of a class around it.

Keep the data private and expose methods[7] to the outside world that would work on the data. This control ensured that fewer errors were made.

---

[7] Don't say function, or you might get lynched. Some people are touchy about the terminology. Methods are for classes. Functions live outside of classes in languages such as C++.

However, Object Oriented Programming is more than just data hiding and code reuse. It's about working with higher levels of order.

4.1.4. <u>Object Oriented Programming - Evolved</u>

Next came next-generation programming languages such as C# and Java. These languages require you to create a class definition before you can begin. Within that class was a main method that was always called first.

Not surprisingly, people (me included) tried to program the old fashion way by putting everything in the main method, but that was awkward.

The language was forcing people to think in a new way – and so the story continued. Each language construct allowed for the greater expression of the Object Oriented paradigm.

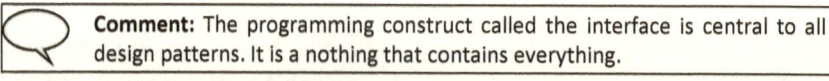

Comment: The intent of Object Oriented Programming is to model a problem domain. Classes and sub-classes encapsulate business objects and their inter-relationships within the problem domain. Different problem domains have different hierarchies of business objects.

We shall visit the language constructs and how they embody the relevant design patterns later.

Let us begin with looking at object oriented programming from a slightly different point of view...

4.1.5. <u>Gang of Four</u>

Since we are discussing Object Oriented Programming, we should mention the Gang of Four (GoF). They wrote the iconic book *Design Patterns: Elements of Reusable Object-Oriented Software.*

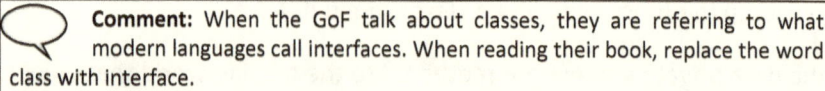

Comment: The programming construct called the interface is central to all design patterns. It is a nothing that contains everything.

Comment: When the GoF talk about classes, they are referring to what modern languages call interfaces. When reading their book, replace the word class with interface.

4.2. Agent Oriented Programming

A dog may be a sub-type of animal,
But Fido is my dog.

"One day an explorer entered a jungle and saw the most amazing thing. The animals were cooperating with each other. Each was doing what only it could do.

"This cooperation wasn't intentional on the part of any animal. It just came about without anyone noticing. The carnivores still ate the herbivores and the herbivores ate the plants. However, everything was happening in a way that made the entire jungle thrive.

"The explorer had seen examples where this didn't happen, where one creature tried to take over and do everything. The perfect example was the corn field. Everything was so uniform and perfect. Then blight came and all was lost.

"A jungle responds to changes by changing the populations within its borders as the stresses on the jungle change. Evolution kicks in and new creatures are born, ensuring greater harmony within the jungle.

From the above story, we see that the jungle doesn't try to do everything by itself. Instead, agents come into being and perform specific tasks.

Each agent or animal knows what it needs to do and does it. Agents don't know how other agents operate and they don't care. In fact, you could replace one agent with another, and as long as it conforms to the same contract, there won't be a disruption in the jungle.

In Agent Oriented Programming (AOP), we create classes of agents to do specific tasks.

We also sub-class each agent when we need to address variations on a theme.

4.2.1. Chess

As an example, say we want to create a chess game.

We have chess pieces and they play on a field, which is the chess board.

The chess pieces normally do the moving on the field, but the field can do some moving and shaking as well.

In chess there are six types of pieces: King, Queen, Bishop, Knight, Rook, and Pawn.

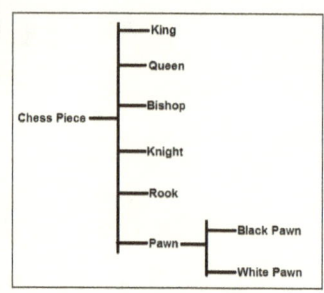

These agents have a certain amount of functionality in common. They have color (black or white), they have position, and they know what a legal move is for them.

For most pieces, color is not important when it comes to functionality. However, the pawn is different in that it changes its behavior as a result of its color. In other words, the pawn has two important {states}. We could use conditional logic to handle this state-based behavior, or we could use {polymorphism}.

To enable polymorphism, we subclass Pawn and then override its base functionality.

When starting a new game, we create an array of chess pieces for each player. When we call the method TryMove, we get back a result that's dependent on the actual piece that was called.

4.2.2. <u>War – An Exercise in Organized Chaos</u>
Let's try another example.

We (the good guys) are going to war against our enemies (the bad guys). What do we do?

The first thing we do is create an army. In this army we have military personal.

What do military personal do? Namely, what do they do to win wars? That depends on the type of soldier.

Let's begin by asking ourselves, what agents do we need to win a war?

We need people to gather intelligence, people to fight the enemy, people to manage supplies, people to heal the wounded and people to manage people, to name a few categories.

In addition, we have personal not directly related to handling the war. For example we have the military police, which are responsible for handling security across the entire military, cross-cutting all other concerns.

Let's display some agents and their major sub-types.

We have six types of military personal listed. Each is a sub-type of the base type 'military personal'. We can always add a seventh sub-type of military personal, should the need arise, but for now, this will do.

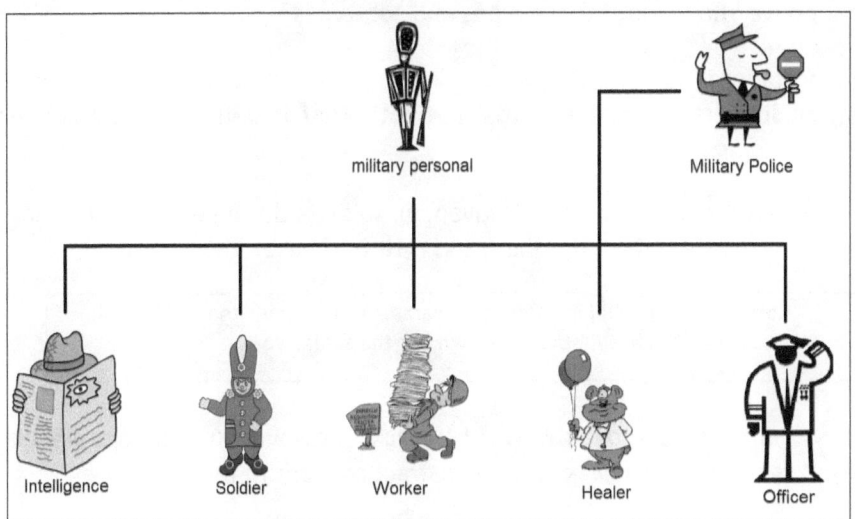

Can we start a war with these categories? No we can't. These are just categories and job titles. What we need are actual people doing their jobs. We need James Bond[8] to do the actual spying. We need Sergeant Slaughter[9] to do the fighting, and so on. When the war program starts, we will create the actual soldiers (objects) to do the work.

Subtypes of Soldier

In the past, an agent with the job title of soldier was good enough. Times have changed and we need more specialized soldiers. It is time to sub-class our soldier class.

If the army (application) was organized correctly, sub-classing the soldier class will have zero impact (in theory) on the rest of the army. This is because all soldiers do the

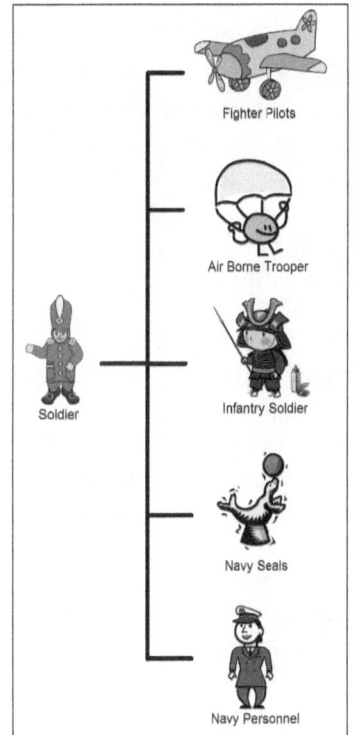

---

[8] The main character in a series of spy movies

[9] A character from the TV series *GI Joe*

same thing. That is, they obey the same contract (they fight).

Here are some variations on the theme of soldier.

What do we know about these types and sub-types?

Polymorphism
First, all soldiers know the words 'Retreat'. That is defined in the soldier class.

When the command to retreat is given, all soldiers do the exact same thing. They stop whatever they are doing and return to base.

 Comment: The Object Oriented paradigm is about the delegation of responsibilities to suitable agents who do the actual work.

Second, all soldiers know the word 'Attack'. However, the generic soldier doesn't know how to attack. It is an abstract concept to them.

So who knows how to attack?

The sub-types of soldier know how to attack:
- The pilot enters the plane, flies to the target and drops bombs.
- The air borne trooper drops onto the target area and engages the enemy.
- Etc.

One command, given to different agents, gives entirely different results for the command 'attack'.

If you will notice, even though all soldiers know how to retreat, different types of soldiers are free to implement their own strategy on how to retreat.

Individual sub-types of soldier are free to override the base implementation of 'retreat', should the need arise.

Sub-types of soldier have to create their own implementation of 'attack', because there is no concrete implementation in the abstract base class of soldier.

Because of polymorphism, the commander doesn't need to know anything about the soldiers, other than that they can carry out the command.

Note: Isn't a navy seal a sub-type of navy personnel?

We can leave the current structure for now and re-factor later, should the need arise. This shouldn't be a problem if the Feature Management team did their job correctly.

Comment: Polymorphism – The ability to treat a subtype as an instance of a base type, allowing the subtype to implement functionality as it sees fit.

This is done by typecasting the subtype as the base type, and then calling methods on the base type. The caller doesn't know the implementation details, and it doesn't need to know.

4.3. Interchangeable Parts

A good programmer is someone who always looks
both ways before crossing a one-way street.
-- Doug Linder --

Object Oriented Programming is all about dealing with various layers of abstraction. We start with code that addresses specific needs and then encapsulate that functionality with higher levels of abstraction.

One of the great powers of OOP is being able to replace one implementation of a class by another implementation of a class. As long as the contract is honored, then all is well.

4.3.1. The Basics
As an example, we need to print a document...

Hard Coding
In the early days of computers, we would attach a printer (Printer123) to our computer. We then create code that knows how to communicate with Printer123 and we print our document.

There is only one problem with this approach. What happens if we change the printer? Doing this would break our code. The code for talking to Printer123 would not work for Printer345.

Encapsulate
The solution is {Encapsulation}. We encapsulate all the functionality for a specific printer within a class definition. The functionality needed to talk to Printer123 would be encapsulated in one class and the functionality needed to talk to Printer345 would be encapsulated in another class.

We then hide the implementation details, to prevent the caller from depending on implementation details.

Adding support for more printers is as simple as adding new printer definitions as needed.

Now all we have to do is create an instance of the class and call the appropriate method to do the printing.

However we still have two problems:
1. The caller needs to know the printer name, so it can use the appropriate class definition.
2. Each printer class definition has its own set of commands.

The first problem can be dealt with using factories and abstract factories (See P. 80, P. 80).

For the second problem we use a contract.

> **Reference:** Encapsulation is the first pillar of Object Oriented Programming.
>
> Encapsulation simplifies caller code by shielding it from unnecessary complexity and allows for more robust caller code. Changes in implementation will not break caller code.

{Contract}
The trick with contracts is that we define a set of commands that all implementations of printer classes are required to follow.

Since all implementations implement the same commands, we don't need to know which class we are instantiating. The result will be the same. The command will go to the correct printer.

There are three ways to implement the contract:
1. Just create the class definition, and name all functions as appropriate. This works best with dynamic languages.
2. Inherit from a base class.
3. Inherit from an interface.

Point 1 can be ignored, since it's not an object oriented way of doing things, and it is error prone.

{Inheritance} from a base class
Inheritance just means we start with existing functionality and add or modify as needed. However we *Never* change the original implementation.

```
namespace PrintBase
{
    /// <summary>
    /// The Printer class is closed for modification, but open to extension,
    /// through the use of subclasses (Open-Close principle of S.O.L.I.D.
    /// design.
    /// </summary>
    public abstract class Printer
    {
        // Implements the print function.
        // It can be overridden in a child class
        public virtual void Print()
        {
            // Do something
        }

        // Must be implemented in a child class
        public abstract bool IsReady();
    }

    public class Printer123 : Printer
    {
        public override void Print()
        {
            // Perform some action
        }

        public override bool IsReady()
        {
            // Perform some action
        }
    }
}
```

{Inheritance} from a contract (Interface construct)

With pure contracts or interfaces we have to supply all functionality. We can't rely on existing functionality.

Of course we could overcome this by using composition. Instead of writing all functionality, the function could just call a library function.

```
namespace PrintBase
{
    public interface IPrinter
    {
        public virtual void Print();
        public abstract bool IsReady();
    }

    public class Printer123: IPrinter
    {
        public override void Print()
        {
            // Perform some action
        }

        public override bool IsReady()
```

```
        {
            // Perform some action
        }
    }
}
```

Now we have potentially dozens of implementations for dealing with all our printers, and a consistent way to invoke the functionality. Now the question is, which implementation should we instantiate into an object?

The answer is, we use a factory.

4.3.2. <u>Factory {Factory}</u>

Factory methods encapsulate the process for selecting an appropriate class implementation that fulfills the desired contract.

Let's continue with the example of the printer.

The steps would be as follows:
1. The caller requests a printer object.
2. The factory polls the operating system and gets a list of all available printers.
3. Based on various criteria, the factory returns a printer object.
4. The caller then calls methods defined by the contract.

Obviously the factory does a lot of things. It can even be customized based on the needs of the caller.

However, the caller doesn't know how the factory does what it does. The caller also doesn't know what implementations are available, and it doesn't need to know that.

 Comment: The purpose of a factory is to encapsulate the logic for selecting an implementation that fulfils a specific contract.

```
// The contract is implemented as the interface iPrinter
public void Print(string str)
{
    iPrinter p = PrinterFactory.GetInstance();
    p.Print(str);
}
```

```
// The contract is implemented as the base class Printer
public void Print(string str)
{
    Printer p = PrinterFactory.GetInstance();
    p.Print(str);
}
```

> **Comment:** Factories always return concrete implementations of the contract.

4.3.3. <u>Abstract Factory {Abstract Factory}</u>

In the example above, we created a printer factory that returned classes for handling different types of printers. The returned classes had one thing in common. They either had the same interface, or they derived from the same base class.

With the abstract factory pattern, we are grouping together multiple factories. This allows us to organize the objects returned into a tree structure.

For the above example, let's add file management.

When we want to save an image to file, we call the file factory, and it returns a file object to use.

Combining the two together, we have the final result.

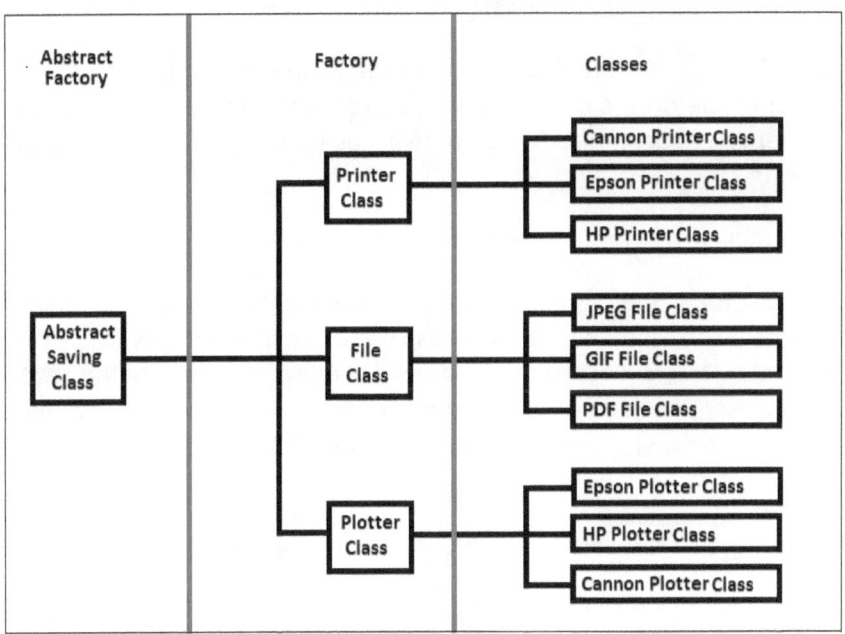

In the above example, the 'Saving' class leverages three other factories to perform its duties.

Depending on the requirements, the abstract factory returns the concrete class appropriate to the situation.

> **Comment:** The abstract factory allows us to organize factories, so that we get the appropriate class for the situation.

4.3.4. <u>Builder {Builder}</u>
Many class objects require a great deal of initialization before they can be of use.

The purpose of the builder is to encapsulate the challenging code required to create the object in one location. This relieves the caller of unnecessary complexity.

As an example, database connectors require special connection strings.

To get the database connector, just supply a few pieces of information to the builder. The builder handles the needful and returns the connector object.

4.3.5. <u>Interpreter {Interpreter}</u>
Before we continue, let's make a slight detour...

Programs are just text documents containing instructions for performing a certain action. Before we can use it, we need to convert it into a format that a computer can understand. There are two way of doing this: compiling and interpreting.

Using a compiler, we create an executable that is delivered to the customer.

Interpreters read the source code and then generate the computer instructions on the fly. Anyone who has created scripts, such as JavaScript, has used interpreters. You open a command window and execute the script. The interpreter reads the script, converts the instructions to machine language on the fly, and executes them.

<u>Embedded Interpreters</u>
Interpreters get interesting when we use them inside applications.

In .NET, we have the CSharpCodeProvider Class. Using this class, we can read a file and create objects on the fly that the developers never created classes for.

4.4. Machine Learning

Setting loose on the battlefield weapons that are able to learn
may be one of the biggest mistakes mankind has ever made.
It could also be one of the last.
-- Richard Forsyth --

The real power of design patterns come into play when we tackle the problem of electronic learning.

In normal programming, we forecast all possible scenarios and create a program that encapsulates these scenarios. Every time a new scenario is encountered, we create a new class, add logic to deal with it and then recompile. Don't forget the need to test. When it comes to artificial intelligence, this is an unworkable scenario.

The alternative is to create a platform that can save and execute scenarios on the fly based on requirements.

The trick here is to make scenarios as granular as possible and use metadata to organize everything.

We then serialize scenarios and store them in a library. When an algorithm is needed, we desterilize the algorithm and use the interpreter to create an object that expresses the functionality needed.

4.5. Working at Scale

I believe in scale Invariance.
I am the equal of a sun or a galaxy,
And a bacterium or an atom is equal to me.
-- Unknown --

A business solution is a hierarchal structure of organized functionality. No one level of organization or order is better or less than another. Each is essential to the functioning of the whole.

Parable:

One day Leonardo Da Vinci visited a construction site. He asked a brick layer what he was doing. With annoyance, the brick layer replied, 'Can't you see? I am laying bricks to build this wall.'

Leonardo Da Vinci visited a second brick layer doing the exact same work and asked the same question. The brick layer replied, "I am building a cathedral.'[10]

---

[10] Unknown author

When faced with a problem, we use functional decomposition to break a problem down into its sub-components. We then organize what we have into a new structure that helps us achieve our goals.

 Comment: The Object Oriented Paradigm is about achieving higher levels of order and organization, in order to fulfill our needs.

However, we can't get to the next level of order unless we master the current level. On the other hand, we will be lost in the woods without a bird's eye view of the terrain.

There are two types of software solutions: Services and Applications.

Services are consumed by other business solutions and don't require a user interface.

Applications are used by people and are deployed on various presentation platforms.

However, when it comes to it, services and applications are pretty much just tools that fulfill human need.

Here are some platforms, in alphabetical order:
- Managed platforms
  - .NET Platform
  - Java Platform
  - Etc.
- OS based platforms
  - Linux OS
  - Mac OS
  - Windows OS
  - Etc.
- Phone based platforms
  - Apple's iOS
  - Google's Android
  - Palm OS
  - RIM's Blackberry
  - Windows Phone
  - Etc.
- Web Browser based platforms
  - Firefox
  - Google Chrome
  - Internet Explorer
  - Safari

o   Etc.

Most business solutions have three tiers of operation. There is a business logic layer, a presentation layer, and some kind of data management layer. Proxies sometimes stand between layers to manage interaction between layers. The following sections will detail the various layers.

5. BUSINESS LAYER

All the world's a stage,
And all the men and women merely players;
They have their exits and their entrances,
And one man in his time plays many parts
-- William Shakespeare, *As You Like It* --

The business layer is the heart and soul of an application. The job of the business layer is to service all approved requests.

Of course the business layer can't operate on its own. It needs a layer through which it talks to clients. It also needs a service layer.

Most applications have all three layers. For simple applications, the layers can be squished together. However, applications that deliver real functionality require the separation in order to improve maintainability and flexibility.

This section will focus on the business layer. The client layer and the service layer will be left for later chapters.

5.1. The Face of Programming

We can never know each other
Except through the masks we wear

In software development, we have at least four ways for agents to do **F.A.C.E.** to **F.A.C.E.** interactions.

5.1.1. {Façade}

As humans, we can only understand and interact with others and the world through the external appearance of thing. The same is true for robots and aliens (unless they are psychic).

Everyone has a face they show to the world. That is a good thing. Otherwise people would see our internal organs and could play with our guts.

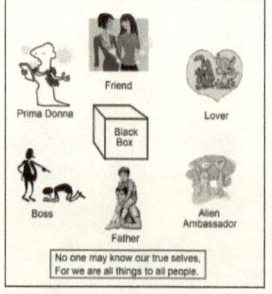

Internal organs are just too complicated and messy to show the world. It is bad for our health and unnecessarily confusing for others.

The same is true for enterprise applications. Their sheer size and complexity makes working with them difficult.

As a result, we create groups of commands (Interfaces) that cater to the needs of a specific agent or user.

This way, the agent or user gets assurances of what to expect when they call methods defined by that contract.

Managing Complexity

The internet is a monster in terms of complexity. Countless protocols exist to perform various functions. Connected to the Internet are trillions of computers talking to each other, each having incompatible operating systems. Even performing the simplest tasks require tons of setup. As a result working with this level of complexity requires centuries of study and training.

At least one would think that.

To deal with this level of complexity, developers identified common scenarios. They then created frameworks to wrap this functionality, exposing methods that cater to the specific needs of the user. The simpler interfaces are used for the common scenario and the advances interfaces are reserved for edge case scenarios.

Do you want to interact with a web service? No problem. Just run a tool and you get a proxy class. You then call methods on that class and interact with it as if it were a local method call.

Do you want to upload a file to a web site? No problem. Just make a call to the appropriate library and don't bother about the detailed interactions needed to pull off that stunt.

Catering to Users

The computer systems of most major banks are monsters. They are designed to cater to millions of clients and can handle millions of interactions a day.

It is foolish to expose that much complexity to all users. It also represents a security risk. As a result the system exposes different façades to different users.

Customers only see functionally appropriate to them. They see their account, services that they can subscribe to based on the type of customer they are, and help based on the context of their needs.

Customer support personal only see functionality necessary for their job and responsibilities. The same is true for every other system user.

Permissions are then controlled by the user group a user belongs to:

- Support
    - Tier 1
    - Tier 2
    - Tier 3
- Customers
    - Free Subscription
    - Basic Subscription
    - Premium Subscription
- Management
    - System
    - Network
    - Website 1
    - Website 2
- Content Creators

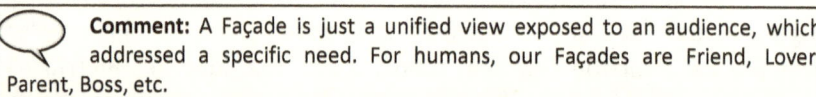

Comment: A Façade is just a unified view exposed to an audience, which addressed a specific need. For humans, our Façades are Friend, Lover, Parent, Boss, etc.

Warning: Contracts must never be too general. Instead they must fit the requirements of the issue at hand. Trying to future-proof an app will only complicate maintenance in the future – unless you are psychic.

5.1.2. **A**dapter {Adapter}

Let's say we are building a news reader and want to subscribe to multiple services for the relevant data. There is only one problem. The format the data returned is different for each service. In addition, the many commands each service understands is different.

With this amount of complexity, the logic for your application will get fragile and hard to maintain.

The solution is to create an adapter class for each service we subscribe to. These adapters cater to the idiosyncrasies of the service, while presenting a single interface for our program to use.

Adding a new service is then as simple as creating a new adapter class and then using it to connect to the service.

 Comment: An adapter is just something that translates messages so both parties can communicate.

Adapters free us from unnecessary complexity by delegate responsibility.

Evolve and Adapt
Adapters are also useful for services that have a tendency to change. From version to version, some services change the way they communicate with their clients.

When this happens you have three choices:
1. Continue to use the same obsolete service
2. Update your code to handle the new service, and then retest to make sure your product is still working correctly
3. Create an adapter class that isolates the changes from your code.

Power to the People
Where did the name adapter come from?

It came from the electrical power industry.

Power supplies can vary from region to region. That means that the hair dryer we bought in the United States will not work in Europe, and vice versa.

Also, the shape of the plugs is different.

The solution here is to buy an adapter that converts the voltage. This allows us to dry our hair anywhere in the world, without having to buy a new hair dryer for every region we visit.

5.1.3. Contract {Bridge}{Contract}
The purpose of the program contract is to define a standardized way of interacting with our service. It defines the services exposed, how to use the services and the format of how responses are returned.

Reference: {Bridge} - Decouple an abstraction from its implementation so that the two can vary independently. [11]

Comment: The {Bridge} pattern describes a many-to-many relationship where an implementation can fulfill multiple contracts, and a contract can be used by multiple implementations.

We use this pattern every time we program to an interface.

Attributes

Attributes are metadata that is used to describe the service. We tag our classes, methods, properties, etc. so that others can know what to expect when they use our service.

By using attribute tags, the caller gains information about a service. The alternative is Documentation! (The problem with documentation that it is a pain to read, hard to find, and can be out of date.)

Attributes are used extensively in many managed environments such as .Net and the Java Runtime Environment.

For example in .Net[12]:

ContextStaticAttribute	Indicates that the value of a static field is unique for a particular context
CLSCompliantAttribute	Indicates whether a program element is compliant with the Common Language Specification (CLS)
FlagsAttribute	Indicates that an enumeration can be treated as a bit field; that is, a set of flags
MTAThreadAttribute	Indicates that the COM threading model for an application is multithreaded apartment (MTA)
NonSerializedAttribute	Indicates that a field of a serializable class should not be serialized
ObsoleteAttribute	Marks the program elements that are no longer in use. This class cannot be inherited

---

[11] Erich Gamma, Richard Helm, Ralph Johnson and John Vlissides, *Design Patterns: Elements of Reusable Object-Oriented Software*
[12] Courtesy: http://msdn.microsoft.com/en-us/library/system.aspx

XML Comments

Xml comments are a great way to annotate your methods. This allows callers to better understand the intent of your code.

Unfortunately only some languages support xml comments. A good example is C# and the .Net framework.

Interface {Mediator}

The interface defines the contract that the service provider guarantees will be met. It defines how we communicate with a service and the format used when returning the result.

The interface simplifies the communication between classes, which is what the Mediator design pattern is about.

In C#, we have:

```csharp
// This interface specifies a contract that allows us to tell
// if two objects are equivalent.
public interface IEquatable<T>
{
    bool Equals(T obj);
}

public class Car : IEquatable<Car>
{
    public string Make {get; set;}
    public string Model { get; set; }
    public string Year { get; set; }

    // Implementation of IEquatable<T> interface
    public bool Equals(Car car)
    {
        return this.Make == car.Make && this.Model == car.Model && this.Year == car.Year;
    }
}
```

> **Comment:** A {Mediator} is any process that simplifies the communication between two classes. In other words, it is just a contract, expressed as an interface.

Library of Moves (Run, Slither, Fly...)

How do you move me? Let me count the ways.

The intent of the bridge pattern here is to separate the implementation of *move* from our class definition.

The easiest way is to create a library of move types and then call on that library whenever you need to do some moving.

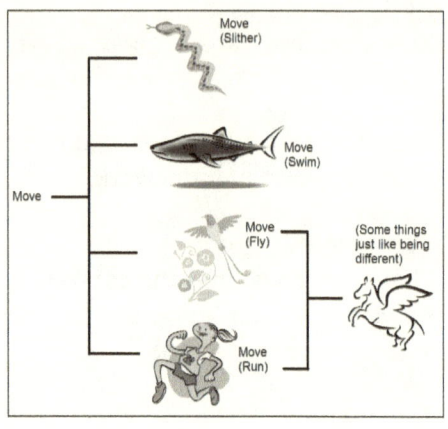

The advantage of placing the move types into a separate library is twofold. First, the same code can be shared by many classes. Second, when a bug is found in the implementation of a method, the fix can be made in one location instead of many locations.

The disadvantage is that changes in the implementation affect all users of the library. This is why we have versioning when dealing with program libraries.

Example

Let's say we want to implement a Human, Snake, Shark and Parrot. We want each to have a default movement type.

The Human, Snake, Shark and Parrot, all derive from Animal. Animal defines an abstract method called DefaultMove.

We could create concrete implementations of this method for each animal type. This is the usual way of doing things.

The only problem is that we will end up with duplicate code if multiple derived classes need the same definition.

We could place the common code in the base class, but that unnecessarily increases the complexity of the base class. Also, we may not have access permission to modify the base code.

The best solution is place the code in a separate library. It makes is easy to change movement types should the need arise, without worrying that our code is correct.

Let us now separate our abstraction from its implementation with an example.

First we have our base class:

```
public class Animal
```

```
{
    public abstract void DefaultMove();

    // Other code
}
```

Next we define our derived classes:

```
public class Human: Animal
{
    // The bridging code. The abstraction (DefaultMove) can now
    // vary independently from the implimentation
    // We can change the abstraction name (DefaultMove to something else)
    // and we cand select another implimentation
    // (change MoveLibrary.Run() to MoveLibrary.Slither())
    public override void DefaultMove() { MoveLibrary.Run(); }

    // Other code
}

public class Snake: Animal
{
    // Actual Implimentation
    public override void DefaultMove() { MoveLibrary.Slither(); }

    // Other code
}

public class Shark: Animal
{
    // Actual Implimentation
    public override void DefaultMove() { MoveLibrary.Swim(); }

    // Other code
}

public class Parrot: Animal
{
    // Actual Implimentation
    public override void DefaultMove() { MoveLibrary.Fly(); }

    // Other code
}
```

Finally we aggregate our variations in a separate static class:

```
public static class MoveLibrary
{
    public static void Slither()
    {
        // Code
    }

    public static void Swim()
    {
        // Code
    }

    public static void Fly()
```

```
    {
        // Code
    }

    public static void Run()
    {
        // Code
    }

    // Other code
}
```

The above code allows both abstractions (Human, etc.) and the implementations to vary independently. Changing the implementation just involves changing one line of code.

5.1.4. Envoy {Proxy}

Many times we need to connect to a service that is not always available. Other times the connection to the service is unreliable, or just slow.

On the other hand, the requested operation could be time-consuming.

Then there is the problem of limited connections. The service is designed to respond to only a few calls at the same time and will reject new requests. An example of this is database connections.

The proxy object connects to the service and requests data. It then caches the data and responds to the application in place of the service.

A second advantage of using a proxy agent is that the agent shields us from any changes in the service API. We just update the class implementation of the proxy agent should the service's API changes.

Of course, with all this sweetness comes some fat.

Storing all that data will increase memory consumption. Also, we are no longer assured that the data we have is up to date.

 Comment: A proxy is an agent that mediates communication between two clients.

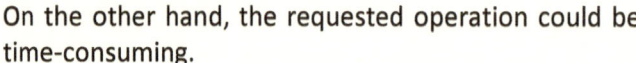 **Comment:** In days of old, the king would send an ambassador to a foreign country. This proxy would represent the king in his dealings with the foreign country. We do the same by sending someone else in out place to a board meeting or party.

5.2. **Behind the Faces**
I've finally learned what 'upward compatible' means.
It means we get to keep all our old mistakes.
-- Dennie van Tassel --

Previously we talked about the faces (interfaces) we used to interact with the different agents in the world. However, contracts and such are useless without something to fulfill the terms of the contracts. To win at cards or anything else, what we need is a bunch of **A.C.E.S.**

5.2.1. **A**ction Methods
Action methods are where the real work takes place. This is where we get down and dirty with less than pretty code. Here is where we communicate with databases and other services. This is where we implement algorithms.

Here are some things we need to keep in mind to manage complexity.

Don't Reinvent the Wheel
When faced with a problem that is non-trivial, check the internet and your information network to see if the problem has already been solved.

Your information network consists of coworkers, company news groups, organizations you subscribe to, and blogs. It also includes the countless web sites dedicated to software development.

Single Responsibility
An action method must do one thing and one thing only.

Action methods that perform multiple tasks vastly increase the complexity of the solution. It makes solution management difficult and the solution less flexible. These action methods are breeding grounds for bugs.

A perfect example of what not to do can be found in Microsoft Office. The redo command changes its behavior when it comes to the end of the redo list. When it comes to the end of the redo list, it changes its behavior to 'repeat previous action'.

One moment, you are redoing your previous actions, and the next minute a serious of letters appears on the screen for no apparent reason. Then you have to undo the previous unintended action – very annoying.

If you need more than one verb to describe what the method does, then you need to re-factor.

Extension Methods {Decorator}

Extension methods, also known as decorators, add new functionality to an existing class definition without modifying the underlying class definition or extending the class.

As an example, the C# string class defines several string manipulation methods. What if we want to extend that class to include additional functionality?

That is impossible in the case of the C# string class, since it is sealed. However, there is a way around this. It is the extension method.

```csharp
/// <summary>
/// String Extension Method
/// </summary>
public static class StringExtensions
{
    public static int WordCount(this String str)
    {
        return str.Split(new char[] { ' ', '.', '?' },
                    StringSplitOptions.RemoveEmptyEntries).Length;
    }
}

/// <summary>
/// Using the extension method
/// </summary>
public class SomeClass
{
    public int SomeMethod(String str)
    {
        // do something

        return str.WordCount();
    }
}
```

Stateless Programming

The philosophy of the stateless design pattern is that you never store any state. A method must only work on data coming in from the input parameters and never uses any class variables to store or retrieve and data.

This greatly simplifies maintenance, since there are fewer dependencies to worry about.

 Comment: The reason programs are hard to debug is that the potential for data evolution is infinite.

You never know what you will get, when the current state of your data is dependent on the previous state of your data. It is even harder to understand if the data can be modified without restrictions.

Rules:
- The method must be marked as *static*
- Never use class variables to store any data.
- Only accept inputs from the method's input argument list
- When necessary, prefer calling stateless methods.
- For .NET programming, use the *Pure* attribute

{Singleton}
There are times when there should only be one instance of a resource.

Reasons include:
- **Security:** There can only be one data repository for security reasons.
- **Cost:** Database providers charge by the number of simultaneous connections provided. By only having a single gatekeeper, we save resources.
- **Time:** Initialization can be time-consuming. The reason is because we need to collect data from multiple resource providers.
- **Resources:** Some services can be resource hogs. On startup they load megabytes or gigabytes or more of data.

```csharp
/// <summary>
/// Singleton class
/// </summary>
public class RareResource
{
    private static RareResource _RareResource;

    /// <summary>
    /// Private constructor. Can't be called externally
    /// </summary>
    private RareResource()
    {
        // Initialize class
    }

    /// <summary>
    /// Caller uses this method to get an instance of the resource class
    /// Note: This method is static
    /// </summary>
    public static RareResource GetInstance()
    {
        if (_RareResource == null)
        {
            _RareResource = new RareResource();
        }
        else
        {
            return _RareResource;
        }
    }
}
```

```
/// <summary>
/// Using the extension method
/// </summary>
public class SomeClass
{
    public void SomeMethod()
    {
        RareResource resource = RareResource. GetInstance();

        // do something with resource
    }
}
```

5.2.2. Cross-cutting Concerns

Cross-Cutting Concerns are aspects of business solutions that touch all aspects of an application.

Concerns include: Security and Logging

Suggested Reading:
Aspect-Oriented Analysis and Design: The Theme Approach,
March 2005 (by Siobhán Clarke, Elisa Baniassad)

Logging
We use logging to keep track of the various transactions performed by a program.

Some reasons for logging are:
- **Auditing**: For legal reasons, sometimes an audit trail is needed when managing user requests. We need to know who did what.
- **Security:** Using this we can tell who interfered with a system. This helps with Repudiation, where anonymous people perform unapproved actions.
- **Performance:** By recording various performance metrics while performing actions, we can identify performance bottlenecks.
- **Maintenance:** Logging can be used to track down hard-to-reproduce bugs. This is often used to collect crash information.

The ideal logging is one where the implemented code focuses only on its own concerns.

Security
All methods that work on user data require a security context before being evoked. A perfect example is the method that needs to contact a database.

Without the proper security context, we might open the way to *Elevation of Privileges* attacks.

The simplest security context is User name and Password.

The ideal security is one where the implemented code has no knowledge of security issues. The code is only executed when the correct security context is passed. Otherwise an exception is thrown.

Parallel Programming

Parallel programming allows us to divide work among several threads. This additional level of organization sits on top of the method that is being executed.

Again, individual pieces of code should not know how it is being used.

Dealing with Cross-Cutting Concerns

There are three ways of dealing with cross-cutting concerns.

The first way is to write the necessary code into all your methods. This is bad programming since it complicates development and maintenance.

Since the code for cross-cutting concerns touch hundreds of places in the code, making sure all those locations are properly updated becomes a nightmare.

The second way to deal with cross-cutting concerns is to have the code call methods at specific locations. This too is a problem, since you need to remember to add the code and knowing where to add the code can be tricky.

The best way to deal with cross-cutting concerns is to ignore them. At compile or packaging time, we run a tool that injects the necessary code.

Code Contracts

Code contracts allow us to define preconditions and post conditions for methods. These conditions appear as attributes we use to decorate our methods. As such, we can safely ignore them when maintaining code.

The great thing about code contracts is that it takes the place of error checking code. This simplifies the development process.

Additionally, code contracts can be used to validate the correctness of a method.

One drawback of code contracts is remembering to add the attributes.

Also, code contracts can impact the performance of code. On the other hand, we can decide the level of enforcement when we produce shippable code.

 Note: System.Diagnostics.Contracts is new to .NET 4.5 – It is available as a download for .NET 4.0
Download at: http://research.microsoft.com/en-us/projects/contracts/

5.2.3. External-Facing Methods

External-Facing methods pull in data from external sources. As such, the data returned from the methods cannot be determined by knowing just the methods arguments.

Data validation is essential for such methods.

5.2.4. State Management Methods (Properties)

Program classes are used to represent business objects and also to store library functions.

Only business objects need to manage state. This state is normally stored in class variables. In the olden days, the only way to make the internal state of the class available to the caller was to make the state public. (You could use methods, but how many programmers did that?)

Unfortunately, this meant that the external caller could place your business object into an invalid state.

To avoid this, language designers created class properties. Externally, these properties looked like class variables, but they have logic to manage state. With the construct below, we can choose to make it read-only by only defining a get. With the set, we can control how the property is changed.

```
/// <summary>
/// Property in C#
/// </summary>
public class SomeClass
{
    // Code

    /// <summary>
    /// Manage state
    /// Note: The private variable is below the property
    /// since they both make one logical unit.
    /// </summary>
```

```
public int State1()
{
    get
    {
        return myState1;
    }
    internal set
    {
        myState1 = DoSomething(value);
    }
}
private int myState1;

public int DoSomething(int val)
{
    // groom input before returning value
}
}
```

5.3. Command and Control (Explicit)

Strategy without tactics is the slowest route to victory.
Tactics without strategy is the noise before defeat.
-- Sun Tzu --

For real-world problems to be solved, we need to make decisions and control how commands are executed. We need to choose the appropriate tactic for the occasion.

The primary concern here is how to take action based on some input. The traditional way is using control and repetition operations. For completeness, here they are.

5.3.1. Decision Making

The decision-making method has only one responsibility. It must select the appropriate tactic based on some input criteria.

If-Else

The If-Else construct is an unstructured but more powerful way of making decisions, since you can control flow using logically different criteria.

However, care should be made to make sure that groupings are consistent. In the example below, we test for animal type (mammal, reptile, etc.), then sub-types (Bats, primates, etc.).

```
/// <summary>
/// Do some cow related action
/// </summary>
public void AnimalAction(Animal animal)
{
    // Check for type of animal
```

```
    if (animal == null)
    {
        NoAnimal();
    }
    else if (animal is Mammal)
    {
        Mammal mammal = animal as Mammal;
        mammal.KeepWarm();

        // Check for type of mammal
        if (mammal is Bat)
        {
            Bat bat = animal as Bat;
            bat.DriveSomeoneBatty();
        }
        else if (mammal is Anteaters)
        {
            Anteaters anteaters = animal as Anteaters;
            anteaters.EatAnts();
        }
        else
        {
            HandleUnknownAnimal();
        }
    }
    else if (animal is Reptile)
    {
        Reptile mammal = animal as Reptile;
        mammal.StayCool();

        // Check for type of reptile
        if (mammal is Aligator)
        {
            Aligator aligator = animal as Aligator;
            aligator.DoTheGator();
        }
        else if (mammal is Snake)
        {
            Snake snake = animal as Snake;
            snake.DoASnakeDance();
        }
        else
        {
            HandleUnknownAnimal ();
        }
    }
    else
    {
        HandleUnknownAnimal();
    }
}
```

Switch-Case
The switch is used for dealing with a finite set of pre-defined choices.

```
/// <summary>
/// Perform action and return if successful
/// </summary>
```

```
public bool PerformAction(ActionType theAction)
{
    switch (theAction)
    {
        case ActionType.Sing:
            return SingSomeSong();

        case ActionType.Dance:
            return DoTheJive();

        case ActionType.Sleep:
            return GoToBed();

        case ActionType.Eat:
            return EatCookies();

        default:
            throw new ApplicationException("Invalid action type");
    }
}
```

5.3.2. Repetition

The purpose of this is to perform a repeated operation as part of an algorithm. Language constructs used to handle data collections will be dealt with later (see p. 126).

For

The 'for' construct controls the number of times an action is performed.

This is used when the number of times an action needs to be performed is pre-determined.

While

The 'while' construct is for when we need to perform an action multiple times, but we don't know beforehand how many times to perform the action.

Do-While

The difference between the 'while' and the 'do-while' constructs is that with the 'do-while' construct, we want to perform the action at least once.

5.4. <u>Command and Control (Implicit)</u>

*Most people have no idea of the giant capacity we can immediately command
when we focus all of our resources on mastering a single area of our lives.*
-- Tony Robbins --

Several design patterns exist that allow us to make decisions without
actually making a decision (Zen-like).

5.4.1. <u>Polymorphism</u>

With polymorphism, the decision-
making action depends on the type
of agent you are controlling.

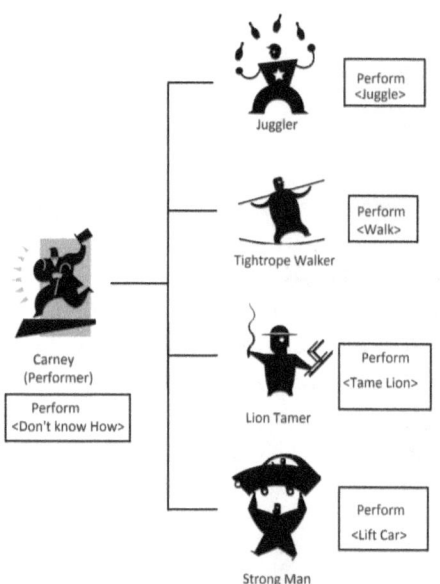

You just collect a group of agents
and give them the same command,
letting them worry about the
implementation details.

Polymorphism is about delegating
responsibility to agents that know
best how to perform an action.
Each agent knows what to do when
given the same command.

<u>Circus</u>
As an example, we are the ring
master in a circus. We have four carneys under us.

We go on stage. For each Carney in our collection we give the command,
'Perform'. The Carney does their trick and exits.

To us, the Carneys are interchangeable. One Carney is as good as another.
As long as the Carney can execute the command 'Perform', we are happy.

```
/// <summary>
/// Perform action. Actual action depends on the implementation
/// details of each Carney
/// </summary>
public void PerformAllActions(List<Carney> carneys)
{
    foreach (var carney in carneys)
    {
        carney.PerformAction();
    }
}
```

Reference: Polymorphism – Treat a child class as an instance of its base class, allowing the child class to override the base class implementation.

Taxonomy - Inheritance

We organize things into groups and sub-groups in the real world.

In our example, we have our Carney. There are things all Carneys know how to do. This is our base class definition.

We then extend our Carney class to take into consideration variations in the theme called Carney. By extending Carney, our new classes inherit the functionality existing in Carney.

With Inheritance we can:
- Override our base implementations. The Lion Tamer overrides the Perform command to perform in a way only a Lion Tamer can.
- Add new functionality that represents the theme of Lion Tamer. In this case we can add the functionality needed to take care of lions. Note: Polymorphism can't here, since it's not defined in the base class.

 Note: Polymorphism should never be used as a way of reusing code. If you want to reuse code, create a library and then reference that library using the {bridge} pattern.

5.4.2. **Divide and conquer {Strategy}**

A strategy is just the implementation of an algorithm, designed to perform a singular task.

We encapsulate an algorithm and expose its functionality through a well defined contract.

In other words, a strategy is just a method within a class.

As an example, we have a math library. This class exposes functions to perform various math operations.

Let's say we need to multiply two matrices. This functionality would be exposed as a strategy (function).

 Comment: When exposing methods as a strategy, it is best to use stateless methods. Otherwise you might get unintended data interactions.

5.4.3. <u>Divide and conquer {Template Method}</u>

Strategies are high-level implementations of an operation. They don't deal with implementation details.

Templates are designed to handle implementation details. For example, the simple *add* operation isn't so simple.

We need *add* functions to deal with the details of data types. For addition we need to deal with:

- Integers
- Floating point numbers
- Complex numbers
- Matrices
- Etc.

Many languages have built-in support for the template design pattern. In C++, it is called a template. In C#, it is called generics.

 Comment: A template is a design pattern that allows us to abstract away the implementation details of a strategy.

<u>C# Generics</u>

C# introduced a type safe way of handling template methods.

When using the generic method, you call the method as usual, passing in a known type. The compiler then generates a method that has the specified type.

```csharp
namespace Namespace1
{
    /// <summary>
    ///  This interface specifies a contract that
    ///  allows us to tell if two objects are equivalent.
    /// </summary>
    public interface IEquatable<T>
    {
        bool Equals(T obj);
    }

    public class Car : IEquatable<Car>
    {
        public string Make { get; set; }
        public string Model { get; set; }
        public string Year { get; set; }

        // Implementation of IEquatable<T> interface
        public bool Equals(Car car)
        {
            return this.Make == car.Make &&
                   this.Model == car.Model &&
```

```
                    this.Year == car.Year;
            }
        }
    }
}
```

5.4.4. Divide and conquer {Chain of Responsibility}

{Chain of responsibility} is about doing what you can and then calling someone else for specialized work. It allows us to simplify the problem by breaking the problem into manageable chunks.

The advantage is that we can simply add functionality as needed without worrying about breaking existing functionality.

```
/// <summary>
/// Perform action and return if successful
/// </summary>
public bool PerformAction(ActionType theAction)
{
    // perform action

    switch (theAction)
    {
        case ActionType.Sing:
            return SingSomeSong();

        case ActionType.Dance:
            return DoTheJive();

        case ActionType.Sleep:
            return GoToBed();

        case ActionType.Eat:
            return EatCookies();

        default:
            throw new ApplicationException("Invalid action type");
    }
}
```

> **Comment:** Isn't it interesting? As developers we use design patterns all the time. However we are seldom aware that we are doing so. However, only through the conscious use of design patters can programming be turned into software engineering.

Delegation

Many times algorithms are almost identical in terms of code. For instance we open a connection to a database, run a custom query, get back the results and then perform a custom operation on all records returned.

Normally we would need to create multiple functions to handle each implementation. This introduces the possibility for mistakes.

To overcome this, we create one method with all common functionality (DoAction). We then invoke that method and pass in a reference to a custom strategy (Strategy1). Changing the strategy method changes how the code behaves at runtime.

```csharp
public void SomeFunction()
{
    int [] arr = new int[10];

    // We invoke the configurable function to do something.
    // We pass in a strategy object to perform the custom action
    DoAction(arr, Strategy1);
}

/// <summary>
/// Chain of responsibility. We do what we can,
/// and then let someone else do what only they can
/// </summary>
public void DoAction(int [] arr, Action<int> action)
{
    // Do setup

    foreach (int val in arr)
    {
        // Delegate responsibility to guest worker
        action.Invoke(val);
    }

    // Do cleanup
}

public void Strategy1(int input1)
{
    // Do something
}
```

Comment: In the above code, *Action* is a C# program construct with the innovative name of Delegate. It allows us to delegate responsibility by passing in a strategy object. When called, the strategy does the needful.

A little bit of history:
The concept of operation delegation started with function pointers in C++. This allowed functions to call other functions dynamically at runtime. The only problem was that there was no assurance of what we would get.

Fast forward and we have .NET 2.0. With .NET 2.0, we have program delegates. It allowed us to wrap a method call within a type-safe wrapper. This made interacting with unknown methods more manageable.

Delegation is most useful with complex algorithms where multiple actions need to be performed before and after a custom step.

Recursion

A specialized case of delegation is recursion. With recursion, we call ourselves with a simpler representation of the problem, until the solution becomes trivial.

```
int NthFact(int n)
{
    if (n <= 0) throw new ArgumentException();
    if (n == 0)
    {
        // Base case
        return 1;
    }
    else
    {
        // Calls itself with simpler problems
        return n * NthFact(n - 1);
    }
}
```

6. DATA LAYER

Intuition becomes increasingly valuable in the new information society
precisely because there is so much data.
-- John Naisbitt --

The data layer has one responsibility. It takes care of the data needs of the business layer.

6.1. Class Objects

Experts often possess more data than judgment.
-- Colin Powell --

A class definition represents a *singular* noun in the business space of the solution. It defines the properties of the noun and its capabilities (methods).

However, for a class definition to be useful it must be instantiated into an object. That is, it must be turned into so something that can store data and perform tasks. After all, the design plans for a lawn mower will not cut the grass.

Here are some basics on object and variable management.

 Comment: A Person is a singular object that represents the attributes of a person. A List of Person objects is also a singular object that stores multiple instances of Person.

6.1.1. Object Creation

When a class definition is turned into an object, sufficient space is allocated on the heap for that object.

This space is just a consecutive block of memory used for storing the actual values defined in the class.

The actual values are called class variables.

The 'new' keyword is required when an object is created from a class definition. This keyword tells the system to allocate the required memory, and then returns the address of the object.

Comment: The heap is just an unstructured area for storing objects. Think of heap as a heap of clothes.

Class variables

Class variables are just the variables defined at the class level. They live outside of class methods.

Space for class variables is allocated at object creation time and exists for the lifetime of the object. As a result, they don't require the 'new' keyword.

Primitive data types include:

- Integers
- Floating point numbers
- Booleans
- ❖ Pointers (discussed later)

The actual values of the variables are undefined at object creation.

Most languages require you to define the value of a variable before reading its data. With C this is not the case. As a result, C variables can take on arbitrary values unless explicitly initialized.

 Comment: Value type data is called variables. They don't require the keyword 'new' to exist. They are in existence the moment the class they are embedded in is turned into an object, or the method they live in is entered. However, variables must be initialized, or their value will remain undefined.

Custom Value types (struct)

Custom value types can be defined. They are just a composition of multiple primitive data types. Just like primitive data types, they too live within the space allocated for the object.

In C#, custom value types are defined using the keyword 'struct'. They are defined in a similar way as classes.

For historical reasons, C# structs can't inherit from other structs. They however enjoy the benefits of encapsulation and can define methods and properties.

 Comment: In C#, all value types are defined as structs.

 Comment: In C#, a property is a managed wrapper that controls access to the data encapsulated within a class or struct. Externally, it looks like a variable. Internally, it defines get and set methods to control data access.

Assignment
The assignment operator copies the value of a variable or the return value of an operation. It then assigns it to a variable.

In the example below:
1. CheeseAge1 has the original value of 5
2. CheeseAge2 has the original value of 2.
3. In the constructor:
    a. CheeseAge2 is assigned the value contained in CheeseAge1.
    b. After the assignment, CheeseAge2 gets the value of 5.
    c. This value will remain, even if CheeseAge1 changes.

In the UpdateCheese() method:
1. CheeseAge1 is assigned a new value, causing the original value to be lost.
2. For the command newClass1 = new Cheese1();
    a. The 'new' operator creates an object of type Cheese1, and then returns a reference to the location where it is stored in the heap.
    b. The assignment operator overwrites the value in newClass1 with the new value.
3. Finally newClass2 is assigned the value of newClass1. In other words, it now points to the object that was created in step 2.

Since both newClass1 and newClass2 point to the same object, we can manipulate both objects using either variable.

```
class Cheese1
{
    int CheeseAge1 = 5;
    int CheeseAge2 = 2;

    Cheese1 newClass1;
    Cheese1 newClass2;

    public Cheese1()
    {
        CheeseAge2 = CheeseAge1;
    }

    public void UpdateCheese(int newAge)
    {
        // CheeseAge1 is assigned a new value, causing the original
        // value to be lost.
        CheeseAge1 = newAge;

        newClass1 = new Cheese1();
```

```
            newClass2 = newClass1;
    }
}
```

Method Variables

When a program executes, a stack object is created. When program control enters a method, variables defined in the method are created and placed on the stack.

When program control enters a new method, more variables are placed on the stack.

When program control exits a method, all variables defined in that method are removed from the stack.

The only difference between method variables and class variables is the scope of operation. Class variables exist for the life of the object. Method variables exist only when program control is in the method. This is of fundamental importance when dealing with multithreading and recursion.

With C/C++, you need to explicitly destroy objects to free the memory. In managed platforms, the objects become eligible for garbage collection when no variables reference them.

 Take Home: The best way to learn object management is by learning C/C++.

6.1.2. Data Access

When creating a class definition, we need to ensure data is accessed in a controlled manner. Occasionally we need to notify observers when data changes.

1. Data Integrity
A class is responsible for the internal consistency of its data. It needs to control who accesses the data and how. It should also work in a multi-threaded environment.

2. Data Access
A data object must control how data is exposed to the world. In C#, this is done through properties defined in class and struct definitions.

3. Data Notification

A data object needs to be able to notify all concerned parties when there is a change to its internal state.

This is done through events. Concerned parties subscribe to an event. They subscribe by registering an event handler. When an event is fired, the agents' event handlers are called, allowing the agents to do whatever they need to do.

6.2. Share and Enjoy {Flyweight}

To write it, it took three months;
to conceive it three minutes;
to collect the data in it all my life.
-- F. Scott Fitzgerald --

The purpose of the flyweight pattern is to create a data object that can be shared among multiple agents.

 Definition: Flyweight is a light weight class in combat sports.

6.2.1. Immutability

An essential feature of flyweight objects is that they be immutable. Once an immutable object is created, it can't be altered. Any operations done on the flyweight object creates a new object.

This is important when dealing with big data applications. With big data applications, multiple instances of the application could be running at the same time, working on the same data set. Immutability allows us to make assurances that we don't have data corruption as a result of timing issues or other failures.

6.2.2. Resource Management

Since flyweights are immutable, we only need to create the flyweight once and then reuse it whenever necessary. This reduces creation time and saves on memory.

Of course, you will need a mechanism to keep track of all the objects you created, so you don't create duplicates. This can be handled using a {Builder}.

Strings

In C#, text strings are objects that can't be modified. Operations performed on strings create new strings.

```
string S1 = "Hello World";

string S2 = "Hello World";
```

Take the above code snippet as an example. Normally the above operation would create two separate objects, encapsulating strings with identical values.

However, C# strings are immutable. This means that we only need one representation in memory. This is possible since there is no danger of the objects changing. S1 will always reference `"Hello World"` regardless of what we do to S2, and vice versa.

6.2.3. No Methods

Flywheels exist only to store data. These objects are passed from one method to another as needed. Therefore having methods goes against the intent of the design pattern, and can cause compatibility issues in the long run.

6.2.4. Serialization {Momento}

Another essential feature of flywheels is that they need to be serializable. Their state must be entirely self-enclosed. If flywheels refer to other objects, then those objects must also be serializable.

In addition, their state must not depend on external resources. It must stand on its own.

> **Comment:** A serializable object has the ability to write its state as a text or binary string ({Momento}). A {Builder} then uses the {Momento} to create a new object (object deserialization).

```
namespace MyFlyWheelApp
{
    /// <summary>
    /// This flywheel depends on another flywheel.
    /// However all of the data is serializable,
    /// and immediately available
    /// </summary>
    public class Flywheel1
    {
        public Flywheel1(int myInt, string myString, Flywheel2 myFlywheel)
        {
            MyInt = myInt;
```

```
            MyString = myString;
            MyFlywheel = myFlywheel;
        }

    public int MyInt { get; }

    public string MyString { get; }

    public Flywheel2 MyFlywheel { get; }
    }
/// <summary>
/// This flywheel has no external dependencies
/// </summary>
public class Flywheel2
{
    public Flywheel2(int someInt, string someString)
    {
        SomeInt = someInt;
        SomeString = someString;
    }

    public int SomeInt { get; }

    public string SomeString { get; }
    }
}
```

6.3. Enumerations

Every man builds his world in his own image.
He has the power to choose,
but no power to escape the necessity of choice.
-- Ayn Rand --

Enumerations are named constants we use to define a finite set of choices.

However with enumerations, we are rarely interested in the underlying values. Instead, we are only interested in what they represent.

6.3.1. Traditional Enums

In C#, traditional enumerations are value types derived from integral types such as byte, int, and long.

First we define our Enumeration:

```
public enum ActionType
{
    Sing = 0,
    Dance = 1,
    Sleep = 2,
    Eat = 3
```

```
}
```

This defines all possible actions. It also defines the possible states, or sub-states a system can have.

We then use it to select some action:

```
/// <summary>
/// Perform action and return if successful
/// </summary>
public bool PerformAction(ActionType theAction)
{
    switch (theAction)
    {
        case ActionType.Sing:
            return SingSomeSong("Happy Song");

        case ActionType.Dance:
            return DoTheJive("Do the Jive");

        case ActionType.Sleep:
            return GoToBed("Lullabye");

        case ActionType.Eat:
            return EatCookies("Cookie Monster song");

        default:
            throw new ApplicationException("Invalid action type");
    }
}
```

For completeness, we have:

```
public bool SingSomeSong(string songName)
{
    // Do something
    return true;
}

public bool DoTheJive(string jiveMusic)
{
    // Do something
    return true;
}

public bool GoToBed(string bedTimeSong)
{
    // Do something
    return true;
}

public bool EatCookies(string cookieEatingTune)
{
    // Do something
    return true;
}
```

6.4. <u>Choice Paralysis {Prototype}</u>

We're entering a new world in which data
may be more important than software.
-- Tim O'Reilly --

The number of states an object can be in may not be infinite, but it is huge. Among all possible states, a few states have special meaning to the world. They appear in multiple scenarios and are relevant to the user.

 Comment: The prototype allows us to give the user a pre-defined set of choices.

6.4.1. <u>Want Fries With That?</u>
As an example, we go to a restaurant and order food.

We have two choices:
1.   We can tell the chef exactly what we want. We specify the ingredients and describe how the ingredients should be cooked, and the chef gives it to us.
2.   We choose among a finite number of choices, from a menu.

Each menu item represents a prototype.

6.4.2. <u>Color your World</u>
Colors come in billions of shades, depending on how we mix the primary colors. We can specify a color using its Red, Green, Blue (RGB) values or we can select from a set of predefined values.

In the .Net Framework, we have *System.Windows.Media.Color*. This represents a standard color using 8 values to define a standard color: R, G, B, A, ScR, ScG, ScB, ScA. That's a lot of values. That's why prototypes are useful.

In WPF, *System.Windows.Media.Colors* defines 141 color prototypes. With that many choices, it is easier to just pick than to define our own color.

Color	Hex	Color	Hex	Color	Hex	Color	Hex
AliceBlue	#FFF0F8FF	DarkTurquoise	#FF00CED1	LightSeaGreen	#FF20B2AA	PapayaWhip	#FFFFEFD5
AntiqueWhite	#FFFAEBD7	DarkViolet	#FF9400D3	LightSkyBlue	#FF87CEFA	PeachPuff	#FFFFDAB9
Aqua	#FF00FFFF	DeepPink	#FFFF1493	LightSlateGray	#FF778899	Peru	#FFCD853F
Aquamarine	#FF7FFFD4	DeepSkyBlue	#FF00BFFF	LightSteelBlue	#FFB0C4DE	Pink	#FFFFC0CB
Azure	#FFF0FFFF	DimGray	#FF696969	LightYellow	#FFFFFFE0	Plum	#FFDDA0DD
Beige	#FFF5F5DC	DodgerBlue	#FF1E90FF	Lime	#FF00FF00	PowderBlue	#FFB0E0E6
Bisque	#FFFFE4C4	Firebrick	#FFB22222	LimeGreen	#FF32CD32	Purple	#FF800080
Black	#FF000000	FloralWhite	#FFFFFAF0	Linen	#FFFAF0E6	Red	#FFFF0000
BlanchedAlmond	#FFFFEBCD	ForestGreen	#FF228B22	Magenta	#FFFF00FF	RosyBrown	#FFBC8F8F
Blue	#FF0000FF	Fuchsia	#FFFF00FF	Maroon	#FF800000	RoyalBlue	#FF4169E1
BlueViolet	#FF8A2BE2	Gainsboro	#FFDCDCDC	MediumAquamarine	#FF66CDAA	SaddleBrown	#FF8B4513
Brown	#FFA52A2A	GhostWhite	#FFF8F8FF	MediumBlue	#FF0000CD	Salmon	#FFFA8072
BurlyWood	#FFDEB887	Gold	#FFFFD700	MediumOrchid	#FFBA55D3	SandyBrown	#FFF4A460
CadetBlue	#FF5F9EA0	Goldenrod	#FFDAA520	MediumPurple	#FF9370DB	SeaGreen	#FF2E8B57
Chartreuse	#FF7FFF00	Gray	#FF808080	MediumSeaGreen	#FF3CB371	SeaShell	#FFFFF5EE
Chocolate	#FFD2691E	Green	#FF008000	MediumSlateBlue	#FF7B68EE	Sienna	#FFA0522D
Coral	#FFFF7F50	GreenYellow	#FFADFF2F	MediumSpringGreen	#FF00FA9A	Silver	#FFC0C0C0
CornflowerBlue	#FF6495ED	Honeydew	#FFF0FFF0	MediumTurquoise	#FF48D1CC	SkyBlue	#FF87CEEB
Cornsilk	#FFFFF8DC	HotPink	#FFFF69B4	MediumVioletRed	#FFC71585	SlateBlue	#FF6A5ACD
Crimson	#FFDC143C	IndianRed	#FFCD5C5C	MidnightBlue	#FF191970	SlateGray	#FF708090
Cyan	#FF00FFFF	Indigo	#FF4B0082	MintCream	#FFF5FFFA	Snow	#FFFFFAFA
DarkBlue	#FF00008B	Ivory	#FFFFFFF0	MistyRose	#FFFFE4E1	SpringGreen	#FF00FF7F
DarkCyan	#FF008B8B	Khaki	#FFF0E68C	Moccasin	#FFFFE4B5	SteelBlue	#FF4682B4
DarkGoldenrod	#FFB8860B	Lavender	#FFE6E6FA	NavajoWhite	#FFFFDEAD	Tan	#FFD2B48C
DarkGray	#FFA9A9A9	LavenderBlush	#FFFFF0F5	Navy	#FF000080	Teal	#FF008080
DarkGreen	#FF006400	LawnGreen	#FF7CFC00	OldLace	#FFFDF5E6	Thistle	#FFD8BFD8
DarkKhaki	#FFBDB76B	LemonChiffon	#FFFFFACD	Olive	#FF808000	Tomato	#FFFF6347
DarkMagenta	#FF8B008B	LightBlue	#FFADD8E6	OliveDrab	#FF6B8E23	Transparent	#00FFFFFF
DarkOliveGreen	#FF556B2F	LightCoral	#FFF08080	Orange	#FFFFA500	Turquoise	#FF40E0D0
DarkOrange	#FFFF8C00	LightCyan	#FFE0FFFF	OrangeRed	#FFFF4500	Violet	#FFEE82EE
DarkOrchid	#FF9932CC	LightGoldenrodYellow	#FFFAFAD2	Orchid	#FFDA70D6	Wheat	#FFF5DEB3
DarkRed	#FF8B0000	LightGray	#FFD3D3D3	PaleGoldenrod	#FFEEE8AA	White	#FFFFFFFF
DarkSalmon	#FFE9967A	LightGreen	#FF90EE90	PaleGreen	#FF98FB98	WhiteSmoke	#FFF5F5F5
DarkSeaGreen	#FF8FBC8F	LightPink	#FFFFB6C1	PaleTurquoise	#FFAFEEEE	Yellow	#FFFFFF00
DarkSlateBlue	#FF483D8B	LightSalmon	#FFFFA07A	PaleVioletRed	#FFDB7093	YellowGreen	#FF9ACD32
DarkSlateGray	#FF2F4F4F						

Color Chart[13]

6.5. Prototype Collection

With data collection,
'The sooner the better' is always the best answer.
-- Marissa Mayer --

A prototype collection is just a collection of all prototypes we are interested in.

6.5.1. Traditional Prototype Collection

Developers use prototype collections without realizing it.

System.Windows.Media.Colors is a static class that exposes properties that represent the standardized colors. The user can either define their own custom color or use a predefined color defined by the static class.

---

[13] Chart came from: http://msdn.microsoft.com/en-us/library/system.windows.media.colors.aspx

6.5.2. **Prototype Enums**

Variables can store an unlimited variety of data. However, not all values are valid for the specified domain. That is when enums come into play.

Using the Color example, suppose we want to limit the color choices to five?

We can define an enum containing the five predefined color prototypes.

For general value types, we have (Not supported in C#):

```
/// <summary>
/// Define an enumerated list of color objects, where only these colors are
allowed
///
/// </remarks>
public enum CarColors : System.Windows.Media.Color
{
    Red = Colors.Red,
    Black = Colors.Black,
    Beige = Colors.Beige,
    Blue = Colors.Blue,
    White = Colors.White
}
```

We can then use it when we want to limit our color choices to only the above five colors:

```
/// <summary>
/// Perform action and return if successful
/// </summary>
public void ColorMyCar(CarColors carColor = CarColors.White)
{
    // Set car color
    This.CarColor = carColor;

    // Do other actions
}
```

 Comment: C# doesn't support this type of enum. This is just an example of what should be supported.

6.5.3. **Delegate Enums**

When dealing with a finite set of choices, we use a switch-case block:

```
{
    public class MyClass
    {
        public void DoSomething(ActionType someAction)
        {
            switch (someAction)
            {
                case ActionType.Dance:
                    // Do something
```

```
                    break;

            case ActionType.Eat:
                // Do something
                break;

            case ActionType.GoToBed;
                // Do something
                break;

            case ActionType.Sing;
                // Do something
                break;

            default:
                // Do something
                break;
        }
    }
}

public enum ActionType
{
    // Assign a delegate the name 'Song'
    Sing = 1,

    // Do something
    Dance = 2,

    // The name and the method we chose have the same names
    GoToBed = 3,

    // Do something
    Eat = 4
}
}
```

The above is a pain to work with and can be error prone.

If we could use the functionality of an enum, we could have:

```
/// <summary>
/// Standard C# only allows us to extend value types such as int
/// We are extending a delegate type here. The limitation is that
/// the method signatures must be the same.
/// </summary>
/// <remarks>
/// Predefined Delegates:
/// - Action: http://msdn.microsoft.com/en-us/library/system.action.aspx
/// - Funct:  http://msdn.microsoft.com/en-us/library/bb534960.aspx
/// - Predicate: http://msdn.microsoft.com/en-us/library/bfcke1bz.aspx
/// </remarks>
public enum ActionType : Func<string, bool>
{
    // Assign a delegate the name 'Song'
    Sing = SingSomeSong,
```

```
    // Do something
    Dance = DoTheJive,

    // The name and the method we chose have the same names
    GoToBed,

    // Do something
    Eat = EatCookies
}
```

We would then use it like so:

```
/// <summary>
/// Perform action and return if successful
/// The default parameter specifies a dance action
/// </summary>
public bool PerformAction(string songName, ActionType theAction =
ActionType.Dance)
{
    // We perform the action and return the result
    return theAction(songName);
}
```

This allows us to encapsulate the logic in a way that allows us to code once, and use multiple times.

6.6.   <u>Be like Water {State}</u>

Snow is water,
and ice is water,
and ~~water~~ (steam) is water,
these three are one.
-- Joseph Dare --

The state of an object refers to the properties of an object that controls its behavior.

Some types of objects change their behavior when they change their state. A perfect example is water.

When the temperature of water is raised above the boiling point, it becomes steam. When below freezing, water becomes solid. The water is the same, but its state is different.

In the same way, the behavior of a solution can vastly change, based on some change in the system.

For instance, a system will respond to different users in different ways. Also, the UI can change based on the properties of a display. The UI exposed on a cell phone will be different than one designed for a desktop computer.

6.7. Data Collections

Data is not information,
information is not knowledge,
knowledge is not understanding,
understanding is not wisdom.
-- Clifford Stoll --

Collections of data come in three main forms: data streams, data arrays, and data graphs.

Just like data objects, data collections need to be encapsulated to maintain data integrity.

6.7.1. Streams

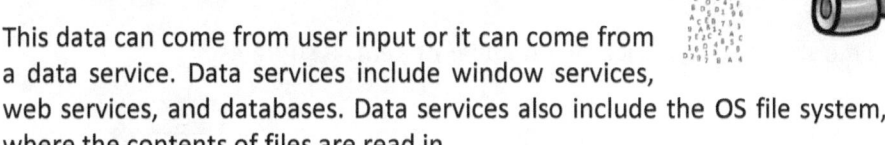

A data stream a sequential input of data from any number of sources. With data streams, you never know how much data will eventually arrive.

This data can come from user input or it can come from a data service. Data services include window services, web services, and databases. Data services also include the OS file system, where the contents of files are read in.

Data streams can also originate from within the application. This includes application resources accessed at runtime.

6.7.2. Arrays

An array of data is just data stored sequentially in memory. Since the data is physically ordered in fixed sized buckets, access is done in linear time. However, reordering the data means physically moving the data around.

Data arrays have a fixed amount of space allocated to them. That means that we need to copy the existing data to a new location, should more data need to be added.

 Comment: A C# List<T> is just an encapsulated array. It manages the array, so we don't need to.

6.7.3. Graphs

Data graphs are data objects linked together in various ways. The business needs of the solution dictate how the data is organized.

The most common ways of organizing data are:

Linked List

With a linked list, we have a 'head' element. This element connects to the next element with a 'next' link, which then connects to the next element, like beads on a string.

Data is accessed in linear time, since we need to traverse the list once.

Stack Class

A stack class has two main operations: Push and pop.

'Push' places data onto the top of the stack. 'Pop' removes data from the top of the stack.

Stack of Plates

The philosophy here is: First in, First out (FIFO).

A useful operation is 'Peep'. This allows you to see the top object without removing it.

Some stacks keep track of the number of items it contains.

Queue

A queue is a linked list where you place items at the 'tail' and remove items from the 'head' of the queue.

Data Tree

The most famous data structure is the Binary Search Tree (BST). Each node has a maximum of two child nodes.

First Come, First Served

In the general case, each node can have multiple children nodes and leaves.

Nodes and Leaves

When dealing with data trees, we can have two kinds of items in the tree: Leaf nodes and Branch nodes.

Branch nodes can have child nodes. Leaf nodes are found at the end of branches and can't have child nodes. In general, a branch node is of a

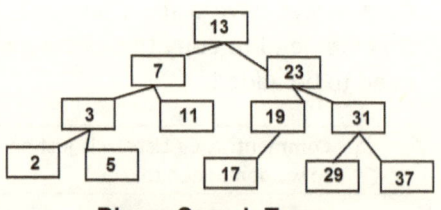

Binary Search Tree

different type than a leaf node. That's not surmising. Branch nodes usually represent containers and the leaf folders represent the content.

However, a branch is just a type of leaf.

6.7.4. <u>Leaves are Just Nodes {Composite}</u>

According to the {Composite} design-pattern, you should treat branches and leaves alike. This simplifies working with the data structure.

Let's take the OS file system. We have files, representing leaves and we have folders, representing branches.

At first glance they seem different, requiring different APIs. However they are similar.

They have:
- Names
- Icon
- Properties
- Command actions

Depending on the file type, the OS displays a different set of commands for the user to use. This holds true for folders.

A folder can be viewed as a special type of file, with its own set of commands.

For instance, when we select 'Open' on a folder, it executes the appropriate command (through polymorphism). The same is true when a file was selected.

6.8. <u>Data management {Visitor}</u>

The goal is to transform data into information,
and information into insight
-- Carly Fiorina --

Our application has to manage collections of data objects and streams of data objects. The best agent for the job is the visitor agent.

The visitor visits all the elements of our collection of data and performs an action based on the type of data found.

Of course, the internals of the data collection are hidden from view. That is a good thing, since the way data is organized differs based on the type of collection we need to deal with.

So how do we visit our data to do our jobs? The answer is that the data collection object exposes a method for stepping through the data. This is the enumerator.

6.8.1. Enumeration {Iterator}

The purpose the enumerator is to allow a {Visitor} to step through all the data in the collection, without worrying about how the data is organized.

The interface in C# is IEnumerable. This returns an IEnumerator object, which allows us to step through the data.

Some collections expose a property that tells us how many items there are in the collection. When this is the case, we can use the 'for' operation.

For

The 'for' construct is used when we know the total number of data items there are in the collection, and we want to go through the entire collection from beginning to end.

This only works if the collection can be accessed by index. Also, it should only be used if the elements can be accessed in constant time.

```
int[] arr = { 1, 2, 3 };
int count = 0;
for (int i = 0; i < arr.Length; i++)
{
    // Constant time operation.
    count += arr[i];
}
```

While

The 'while' construct is used when we either don't know how many data items there are in the collection, or we want to stop before going to the end of the collection. The reason for stopping could be because we found what we were looking for.

The example below uses indexing. We can do this if:
- The number of items in the collection is constant and known
- Access through indexing is linear

```
int[] arr = { 1, 2, 3 };
int count = 0;
int i = 0;
while (i < arr.Length)
{
```

```
        // Constant time operation.
        count += arr[i];
        i++;
}
```

The next example uses an enumerator. Indexing isn't used, so the length can be undefined.

```
List<int> arr = new List<int>() { 1, 2, 3 };
int count = 0;
var enumerator = arr.GetEnumerator();
while (enumerator.MoveNext())
{
        count += enumerator.Current;
}
```

Do-While
The do-while construct assumes a collection has at least one element. If this is not the case, then this will fail.

```
int[] arr = { 1, 2, 3 };
int count = 0;
int i = 0;
do
{
        // Perform action, then check repeat criteria
        // Constant time operation.
        count += arr[i];
        i++;
} while (i < arr.Length);
```

Foreach {Iterator}
In C#, the foreach command encapsulates the 'while' logic, to make a more straight forward {visitor} mechanism.

The foreach is limited in that:
1. Each element is {visited} only once.
2. Elements can't be added or removed from the collection.
3. The order of the collection cannot be changed.

This works best for streams of data with unknown length.

```
List<int> arr = new List<int>() { 1, 2, 3 };
int count = 0;
foreach (var item in arr)
{
        count += enumerator.Current;
```

}

6.9. <u>Data Storage</u>

The new SAS and SATA drives
will change key storage dynamics,
particularly in backup and recovery.
-- Arun Taneja --

Data storage is essential for most solutions. We need to store and retrieve the data needed by customers. We also need to store configuration data as well as auditing and log data.

6.9.1. <u>Serialization {Memento}</u>

The purpose of serialization is to create a string representation of the data of an object. With this string, we can reproduce the object in another place or time.

This is fundamentally important in distributed systems. This allows us to recover from crashes, store information in databases, and pass data throughout the internet.

Another useful implementation of the {Momento} is the undo list for a text editing app. A change in {State} of the document triggers a save operation. Then an undo or redo operation is performed to shift through the saved states.

6.9.2. <u>Data Files</u>

In the beginning of the computer era, all data was stored in files. Even now, we store data in files. Common examples include: XML files, JSON files, Excel spreadsheets, and text files.

For some solutions, data files are perfect. Examples of this are files produced by office suits such as Microsoft Office and Apache Open Office.

For storing configuration data, an excellent solution is the XML file or JSON file.

6.9.3. <u>SQL Databases</u>

SQL databases have been the traditional way of storing and manipulating large quantities of data. When hosted on dedicated servers, they enable

data warehouses that serve the needs of some of the largest companies in the world.

SQL (Pronounced sequel) databases store data in tables. The data is then retrieved and manipulated using Structured Query Language (SQL).

SQL Queries
Structured Query Language was created to manipulate the database, and store and retrieve data.

To add or retrieve data, just send an ad-hoc query to the database.

Unfortunately ad-hoc queries are bad for two reasons:
1. The database has to create a query plan every time it receives a query. This is time-consuming and greatly slows down the database.
2. SQL queries represent a security vulnerability. A hacker could potentially hijack the data connection and get unlimited access to the data. After all, the ad-hoc query is nothing more than a text string that is interpreted at the destination.

 Reference: A good reference to learn SQL is:
https://www.w3schools.com/sql/default.asp

Stored Procedures
Stored procedures are the compiled version of ad-hoc queries.

Their advantages are:
1. Stored procedures are fast. I had previously used SQL queries to generate SQL reports. At one point, the queries took almost ten seconds to run. Then I tried a stored procedure. I got the result in less than a second.
2. Safer data. Strings are always interpreted as strings, and not hidden SQL commands from potential hackers.
3. More maintainable code. By using only stored procedures, we can change the SQL code, and if the interface is the same, no one will know the difference.

 Warning: An application must only access a database through stored procedures.

 Warning: Stripping away potential keywords may sound like a good idea. However that is not guaranteed safe. Also, doing so will compromise the quality of the user experience.

Comment: Stored procedures give us both security and performance, while increasing the manageability of a solution.

6.9.4. NO SQL Databases

Over the decades multiple data storage solutions have come into existence. These are all grouped under the name NoSQL, as in Not SQL. Others call it Not Only SQL.

Here are a few database types in use.

Document Database
Document databases are optimized to store data in standard data formats.

Text formats include: XML, JSON, YAML (Yet another markup language)

Binary formats include: BSON (Binary JSON)

See: https://en.wikipedia.org/wiki/Document-oriented_database

Key-Value Databases
The databases stores key-value pairs, allowing for lookup of data in near constant time.

See: https://en.wikipedia.org/wiki/Key-value_database

Graph Database
Graph databases allow you to represent data in terms of their relationships.

For more details, see: https://neo4j.com/developer/graph-database/

 Comment: The name NoSQL started out as a twitter hash tag.

6.10. Data Connection {Proxy}
You can have data without information,
but you cannot have information without data.
-- Daniel Keys Moran --

Our program needs to connect to various data services in order to function properly. These services include web services and database connections.

The problem with these services is that they could publish newer versions of their services. You then have a choice. Either use the current service, even though it is obsolete, or update your application.

Neither solution is a good one.

The third solution is to use a proxy to represent the service you are connecting with.

There are two advantages with using a proxy.

1.   It isolates changes in the service.

The proxy allows the service code and your solution code to vary independently, allowing for more flexible code.

2.   Data buffering

A proxy can buffer data, thereby reducing latency and bandwidth usage.

7. PRESENTATION LAYER

As both a fine artist and a graphic designer,
I specialize in the visual presentation of words.
-- John Langdon --

Our work is the presentation of our capabilities.
-- Edward Gibbon --

The presentation layer has two responsibilities. First, it displays information to the user. Second, it responds to commands from the user.

7.1. Ancient History

In ancient times, history was made.

In the good old days of software engineering, the presentation of data was deeply tied to the manipulation of data and the business logic.

First you get a command, and then you performed some action based on the command. Since this was done through command line programs, there was no real user interface. You just requested user data whenever you needed it.

Then came the Graphical User Interface...

As user interface technology evolved, people realized that even small changes to the user interface caused breaking changes to the program.

As a result, design patterns such as {Model-View-Controller} (MVC) for web applications, {Model-View-Viewmodel} (MVVM) for desktop applications, and others came into existence. People began separating their presentation logic from their business logic.

This allowed for a more stable design. However, this solution wasn't perfect. The business logic was safe, but the program could still break if a single button on the input screen was renamed.

> **Comment:** Windows Presentation Foundation (WPF) uses the Model-View-Viewmodel (MVVM) design pattern to create feature-full applications that adapt to varying form factors and platforms.

> **Reference:** "In days past children used to play with paper dolls. The paper dolls came with several outfits that could be placed over the doll to dress it up. Software skins do the same thing for the graphic user interface (GUI) of software programs, that paper doll dresses did for paper dolls. Applying a new skin to a program changes the color scheme, theme or style, buttons and controls. It gives the program a fresh look."[14]

7.1.1. <u>User Interface (UI)</u>

The user interface (UI) of an application is just the {façade} that the application presents to the user. Its purpose is to present information to the user and accept user input.

Being a façade, the UI can change depending on the needs of the user. For instance, the application will present a different UI depending on who the user is. An admin will see one UI while an end-user will see another UI.

Using different façades for different users serves two separate functions:

1. It simplifies the UI by removing elements that the user doesn't have permission to use.
2. It simplifies security, since only commands and information appropriate to the security context are displayed.

Let's start with UI Basics.

7.2. <u>Layout Elements</u>

Practice safe design: Use a concept
-- Petrula Vrontikis --

Good design keeps the user happy,
the manufacturer in the black
and the aesthete unoffended
-- Raymond Loewy--

User interfaces have visible elements common to all user interfaces, although few applications use all the elements. They can be grouped into families and sub-families. This allows developers to simplify the development process.

It also simplifies user interaction, since it allows using previous experience to understand new features.

---

[14] Comment taken from: http://www.wisegeek.com/what-are-software-skins.htm

7.2.1. <u>User Windows</u>

Content is exposed primarily through various windows. In the beginning, this was done through the command line. That was fine for developers. However, only the Graphical User Interface could make applications accessible to a larger audience.

Command Line

Command line applications don't have a graphical user interface. Instead the user types commands at the command prompt.

This method of control is popular with many developers and system administrators. The most important advantage of command line applications is that they can be controlled by command scripts to do predefined tasks.

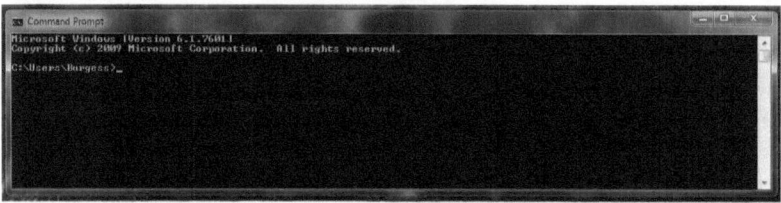

Single Document Interface (SDI)

The single document interface is designed to allow the user to edit only one file at a time. Sometimes developers use this model since it is simpler to implement.

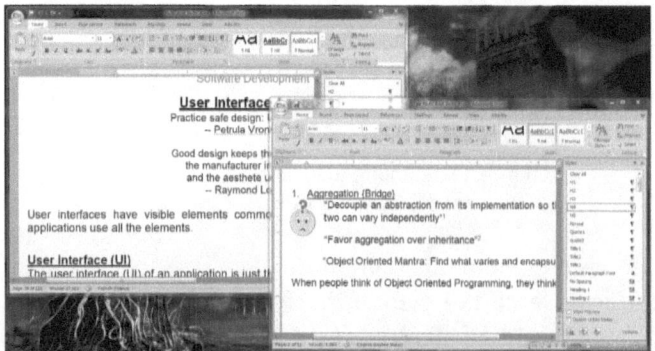

Other times, the SDI is used intestinally to make it easier for the user. Additional instances of the application are created to edit more than one document at a time.

Multiple Document Interface

The Multiple Document Interface (MDI) allows the user to edit multiple documents at the same time. Visual Studio uses the MDI to edit the files that belong in the same solution.

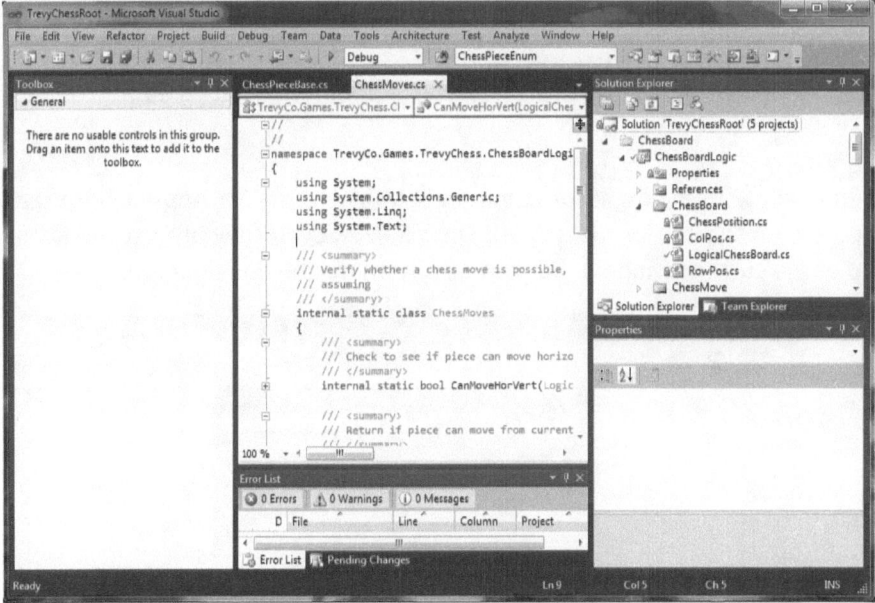

Work Windows

The work window is the place where the main work takes place.

For a simple document editor, it is the area where we compose our text documents. For our favorite Integrated Development Environments (IDEs), they are the windows where we edit our source code.

Tool Windows

In addition to the document window(s), we can have other windows types.

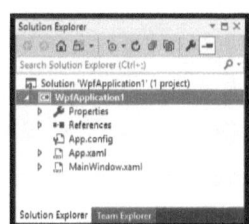

Tool windows house tools in the form of buttons, drop-downs, text inputs and so on.

The search box is a good example of a tool window.

Dialog Boxes

Dialog boxes allow the application to request information from the user, and display messages. An example is the confirmation dialog box.

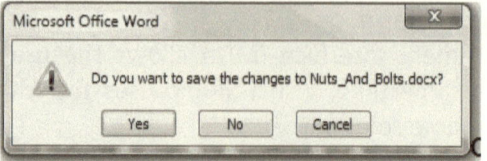

7.2.2. <u>Command Elements</u>

These elements allow users to execute commands or select actions.

Menu

Menus allow us to organize commands in list form. On Apple computers, they are on the top of the screen. On MS Windows, they are usually at the top of the current window, but can be anywhere the developer decides.

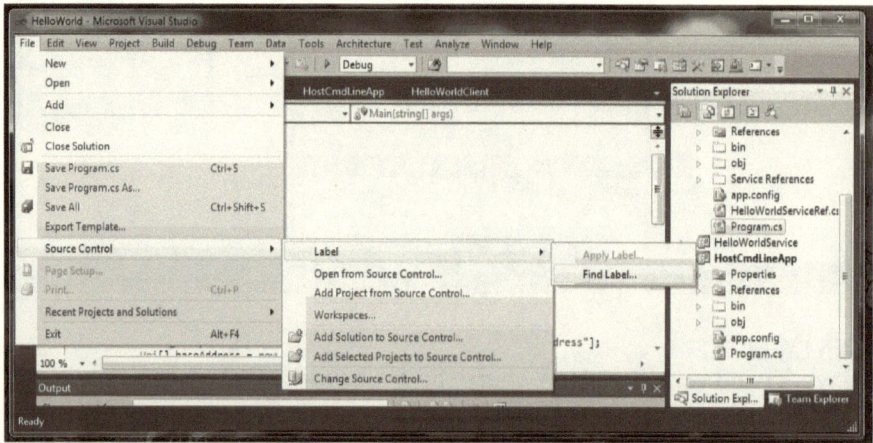

Toolbar

Toolbars allow us to display commands as buttons, Drop-downs, edit boxes, and so on.

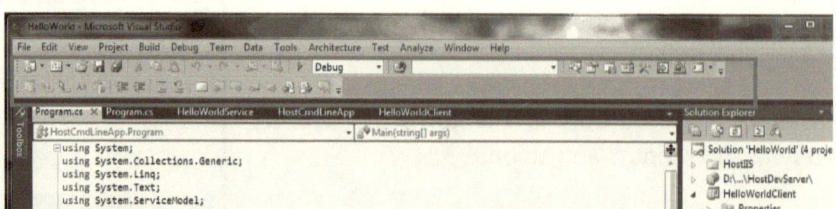

Ribbon Bar

Microsoft designers combined the concept of the menu and toolbar to create the ribbon bar.

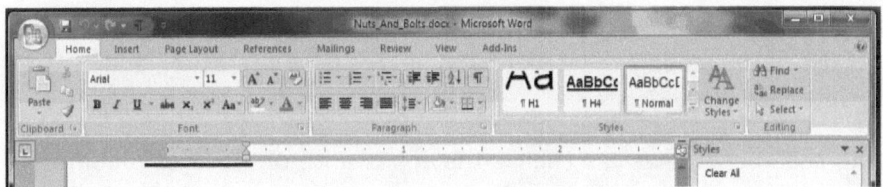

The ribbon has multiple tabs (home, insert, Page Layout, etc.) to organize commands. In addition, the commands in each tab can be organized as well (Font, Paragraph, Style, etc.).

When properly designed, it can make it easier for the user to find commands.

7.2.3. Collection Elements
Another important item is collection items, such as:
- Lists
- Tables
- Drop down boxes

7.2.4. Edit Elements
Edit elements allow you to modify data. These include:
- Text box
- Calendar

7.3. Windows Presentation Foundation
Windows Presentation Foundation (WPF) provides developers
with a unified programming model for building rich
Windows smart client user experiences
that incorporate UI, media, and documents.
-- http://msdn.microsoft.com/en-us/library/ms754130.aspx --

Microsoft's Windows Presentation Foundation (WPF) is designed to create advanced user interfaces for desktop applications.

The user interface is entirely describable using eXtensible Application Markup Language (XAML). Similar to HTML, XAML allows us to define styles. These styles can either be embedded or in an external file.

Comment: XAML is pronounced zamal.

We also have a code-behind file that manages user interaction and controls the interface (view).

With code-behind, we can change the appearance of the UI at runtime and we can respond to user input. This was a vast improvement over existing technologies. It was like plain HTML on steroids. Designers, who knew nothing about programming, could use editing tools to make sexy programs.

Let us begin…

7.3.1. <u>Where is the Main() Man?</u>

Just like windows services, web services, and program libraries, WPF applications don't have a main method. Instead we have an App.xaml file. This file specifies resources, application data, and the startup URI for the application.

> **Comment:** I believe the removal of the main() method is a good thing.
>
> In my opinion, the main() method is a holdover from the Ancient Times when Object Oriented Programming was just an abstract concept.
>
> The main() method allowed people to program in a monolithic way, despite the tools available for Object Oriented programming. Now that crutch is gone.

<u>App.xaml file</u>:

Here we specify the user interface we will display at startup, as well as any resources needed. In this case, the startup view is defined by HellpWorld.xaml.

```
<Application x:Class="HelloWorld.App"

xmlns="http://schemas.microsoft.com/winfx/2006/xaml/presentation"
            xmlns:x="http://schemas.microsoft.com/winfx/2006/xaml"
            StartupUri="MVVM\View\HellpWorld.xaml">
    <Application.Resources>
        <ResourceDictionary>
            <ResourceDictionary.MergedDictionaries>
                <ResourceDictionary Source="Styles\HelloStyle.xaml" />
            </ResourceDictionary.MergedDictionaries>
        </ResourceDictionary>
    </Application.Resources>
</Application>
```

<u>HelloStyle.xaml file</u>:

The style below allows us to define a consistent look and feel across our application.

```
<ResourceDictionary
  xmlns="http://schemas.microsoft.com/winfx/2006/xaml/presentation"
  xmlns:x="http://schemas.microsoft.com/winfx/2006/xaml" >
    <Style TargetType="{x:Type TextBlock}" x:Key="TextStyle">
        <Setter Property="FontFamily" Value="Arial" />
        <Setter Property="FontWeight" Value="Bold" />
        <Setter Property="FontSize" Value="16" />
        <Setter Property="Foreground" Value="ForestGreen" />
        <Setter Property="TextWrapping" Value="Wrap" />
    </Style>
</ResourceDictionary>
```

HellpWorld.xaml file:

This is the file that defines how the user interface should look like.

```
<Window x:Class="Hello.World.SmileyFace"
        xmlns="http://schemas.microsoft.com/winfx/2006/xaml/presentation"
        xmlns:x="http://schemas.microsoft.com/winfx/2006/xaml"
        Title="{Binding Path=HelloWorldTitle}"
        Height="300" Width="300"
        Name="MainWindow1">
    <Grid >
        <Grid.RowDefinitions>
            <RowDefinition Height="*" />
            <RowDefinition Height="5*" />
        </Grid.RowDefinitions>
        <TextBlock Style="{StaticResource ResourceKey=TextStyle}"
Name="HelloMessageText"
                Text="{Binding Path=HelloWorldMessage}"
TextAlignment="Center" />
        <Button Grid.Row="1" Name="MyButton1" Background="White"
Click="MyButton1_Click">
            <Image Name="ButtonImage" Source="{Binding Path=HelloWorldFace}"
/>
        </Button>
    </Grid>
</Window>
```

So far the program is almost completely useless, since it can't respond to user input. We fix that by using events and an appropriate event handler.

7.4. Events {Observer}

The observer, when he seems to himself to be observing a stone,
is really, if physics is to be believed,
observing the effects of the stone upon himself.
-- Bertrand Russell --

The observer pattern, implemented through method wrappers (delegates), events and event handlers, is used whenever we want to keep track of part of the ever-changing electronic world we live in. We request to be notified by the agent whenever it changes one of its states.

In the XAML file above, we specify the event we want to respond to. In this case, it is the Click event of the Button (underlined above). When the user clicks on the button, the method "MyButton1_Click" is called, as shown below.

<u>Sample Code-Behind file</u>:

```
using System.Drawing;
using System.Drawing.Imaging;
using System.IO;
using System.Windows;
using System.Windows.Media;
using System.Windows.Media.Imaging;

namespace HelloWorldApp
{
    public partial class SmileyFace : Window
    {
        public SmileyFace()
        {
            InitializeComponent();
        }
        private void MyButton1_Click(object sender, RoutedEventArgs e)
        {
            ButtonImage.Source =
GetImageSource(HelloWorldApp.Resources.HappyFace);
        }
    }
}
```

The above code gives us:

Before Clicking

After Clicking

Clicking the face causes some action to be performed.

We now have two issues to deal with:
1. How do we handle creating new skins that are simple to manage?
2. How do we localize the view?

7.5. Model-View-Viewmodel {MVVM}[15]

Our work is the presentation of our capabilities.
-- Edward Gibbon --

Introducing the Model-View-Viewmodel (MVVM) design pattern, the next evolution in presentation design for the desktop...

The intent of the MVVM design pattern is to separate what the program displays (view) from the contract that controls the view (Viewmodel), from the data (Model). It is most commonly used with data bound controls.

By separating the three, we can change the view without affecting the rest of the program. Conversely, the view doesn't need to be changed if the {contract} for the viewmodel remains the same.

Here is a simple example of the pattern in action...

7.5.1. Viewmodel {Bridge}

The viewmodel is the contract (interface) that separates the user interface from the business logic, allowing each to vary independently.

When interacting with a user, the application must display certain information. Also, the user must supply information to the application. This is done through properties defined in the contract.

The below code defines the viewmodel of a single User of our system.

```
namespace WpfApplication1
{
    /// <summary>
    /// This is the viewmodel that regulates the interaction
    /// between the Model and the View
    /// </summary>
    public interface IUserViewmodel
    {
        string UserName { get; set; }
        string Name { get; set; }
        string Email { get; set; }
        string PhoneNumber { get; set; }
        RelayCommand SaveContact {get;}
    }
}
```

---

[15] Suggested article, by Jeremy Likness:

http://www.codeproject.com/Articles/100175/Model-View-ViewModel-MVVM-Explained

7.5.2. Model

A data model is similar to a flyweight, except that its data is mutable. (See P. 114 – Share and Enjoy {Flyweight})

The purpose of the model is to store data needed to display to the user or collect data from the user.

There are two types of data: Configuration data and program data.

Configuration data just controls how the UI is set up. This includes user customizations and security context.

Program data includes:
- User info
- Product info
- A single appointment
- A single transaction
- Interaction history
- Etc.

We talked about data in the Class Objects section (P.110).

 Comment: A model is the data that represents a singular static business object. Data models only have properties.

Implemented Model

The data model exposes properties that that bind to the View. It also has a mechanism that tells the view when the data has been updated. In C#, this is done through the INotifyPropertyChanged interface.

Below is an example of the data model. The implementation of the base class can be found in the appendix (Data Mode Base Class, P. 169)

```
//-----------------------------------------------------------
// <copyright file="UserModel.cs" company="CyberFeedForward" >
// Free for use, modification and distribution
// </copyright>
// <Author>
// Trevy Burgess
// </Author>
//-----------------------------------------------------------
namespace TestApp
{
    using CyberFeedForward.WUP.Common.WPF;

    /// <summary>
    /// The data model impliments the viewmodel and has functionality
    /// to save data to user's roaming store.
```

```
/// </summary>
public class UserModel : DataModelBase, IUserViewmodel
{
    public string UserName
    {
        get { return GetState("", SaveType.RoamingSettings); }
        set { SetState(value, SaveType.RoamingSettings); }
    }

    public string Name
    {
        get { return GetState("", SaveType.RoamingSettings); }
        set { SetState(value, SaveType.RoamingSettings); }
    }

    public string Email
    {
        get { return GetState("", SaveType.RoamingSettings); }
        set { SetState(value, SaveType.RoamingSettings); }
    }

    public string PhoneNumber
    {
        get { return GetState("", SaveType.RoamingSettings); }
        set { SetState(value, SaveType.RoamingSettings); }
    }

    public RelayCommand SaveContact
    {
        get { return Command(PerformSaceAction); }
    }

    private void PerformSaceAction()
    {
        // Perform some action when save command is executed
    }
}
}
```

7.5.3. View

Finally we deal with the view.

Code Behind UserView.xaml.cs

The original purpose of the code behind file (below) was to allow developers to separate the business logic from the actual view (XAML file). However, Microsoft didn't go far enough.

If Microsoft had the chance to redesign the system, I'm betting they would toss out this file, since it encourages developers to be lazy. I know I cheat a lot, using this file.

The below code has one purpose. It is to connect the data model with the view, using the command: `DataContext = new UserModel();`

Swapping the View becomes trivial. We can also use an {Inversion of Control} container, such as Unity[16] to swap views on the fly. As long as both the view and the model honor our viewmodel contract, then all is good.

```
namespace TestApp
{
    public sealed partial class UserView : Page
    {
        public UserView()
        {
            InitializeComponent();
            DataContext = new UserModel();
        }
    }
}
```

View file UserView.xaml
The XAML file defines the actual view.

Properties in the view are controlled by data binding. It also uses the {Command} design pattern to notify us of user actions.

In the below code, the magic happens with the binding:
- `Text="{x:Bind User.UserName, Mode=TwoWay}"`

This tells the system what to bind to, and to make sure the data flows both ways.

```
<Page
    x:Class="TestApp.UserView"

xmlns="http://schemas.microsoft.com/winfx/2006/xaml/presentation"
    xmlns:x="http://schemas.microsoft.com/winfx/2006/xaml"
    xmlns:d="http://schemas.microsoft.com/expression/blend/2008"
    xmlns:mc="http://schemas.openxmlformats.org/markup-
compatibility/2006"
    mc:Ignorable="d"
    Background="{ThemeResource
ApplicationPageBackgroundThemeBrush}" Height="478.632"
Width="816.239">
  <Grid>
    <Grid.RowDefinitions>
      <RowDefinition Height="Auto" />
      <RowDefinition Height="Auto" />
      <RowDefinition Height="Auto" />
      <RowDefinition Height="Auto" />
      <RowDefinition Height="Auto" />
      <RowDefinition Height="Auto" />
```

---

[16] For details on Unity, see: https://github.com/unitycontainer/unity

```xml
        <RowDefinition Height="*" />
      </Grid.RowDefinitions>
      <Grid.ColumnDefinitions>
        <ColumnDefinition Width="Auto" />
        <ColumnDefinition Width="*" />
      </Grid.ColumnDefinitions>

    <TextBlock Grid.ColumnSpan="2" Margin="10"
TextDecorations="Underline">Input Data</TextBlock>

    <TextBlock Grid.Row="1" Grid.Column="0" Margin="10"
Text="UserName" FontFamily="Segoe UI"  />
    <TextBox Grid.Row="1" Grid.Column="1" Text="{x:Bind
User.UserName, Mode=TwoWay}" FontFamily="Segoe UI" />

    <TextBlock Grid.Row="2" Grid.Column="0" Margin="10"
Text="Name" />
    <TextBox Grid.Row="2" Grid.Column="1" Text="{x:Bind
User.Name, Mode=TwoWay}"  />

    <TextBlock Grid.Row="3" Grid.Column="0" Margin="10"
Text="Email" />
    <TextBox Grid.Row="3" Grid.Column="1" Text="{x:Bind
User.Email, Mode=TwoWay}" />

    <TextBlock Grid.Row="4" Grid.Column="0" Margin="10"
Text="PhoneNumber" />
    <TextBox Grid.Row="4" Grid.Column="1" Text="{x:Bind
User.PhoneNumber, Mode=TwoWay}" />

    <Button Grid.Row="5" Command="{x:Bind User.SaveContact}"
Margin="10" Content="Save" FontFamily="Segoe UI" />
  </Grid>
</Page>
```

8. FEATURE MANAGEMENT

So much of what we call management consists
in making it difficult for people to work.
-- Peter Drucker --

It is impossible to define a feature with 100% accuracy, since any definition we come up with ultimately depends on the human language to express it.

The purpose of feature management isn't to define features with 100% accuracy, but with enough accuracy to satisfy our customer's needs. Anything beyond that is a waste of time and resources.

8.1.   The Language of Requirements

To be stupid, selfish, and have good health
are three requirements for happiness,
though if stupidity is lacking, all is lost.
-- Gustave Flaubert --

Before a product is created, we need to know what we want. If it is a commercial product, then it is necessary to write down the requirements so that all the stakeholders understand what is needed.

For instance, the programmer needs to know what to build. The tester needs to know if the product is working according to the needs of the customer. The manager needs to know whether or not the feature meets the customer's needs. The customer needs to know how to use the product.

To satisfy these needs, various documents are created. This includes test plans, development plans, and product documentation. No matter the documentation produced, certain rules must be followed.

8.1.1. International English

For better or worse, English is the language of science, engineering, commerce and diplomacy.

Unfortunately, English has countless dialects and local variations. That means one word can mean different things to different people.

International English Dictionary

Fortunately, most people who claim to speak English understand International English.

International English is the language that diplomats use. It is also the language international news reporters use when giving reports intended for a global audience.

Therefore, it is a good idea to describe all requirements using international English – Unless the entire team is composed entirely of local employees.

8.1.2. <u>Avoid Jargon</u>

Jargon is language that is specific to your organization and isn't widely recognized by the engineering community.

It's best to avoid urban slang or jargon, since that makes it easier for new team members to get up to speed with our product. It also helps customers satisfy their needs.

This doesn't mean you can't use language specific to your audience. All you have to do is create a dictionary that everyone can agree on.

8.1.3. <u>Proof Read</u>

All requirements documentation must be proof-read by at least one person not related to the project. This prevents hidden assumptions from getting in the way of clearly stated requirements.

Ideally, the proofreading and correction process will be done in a dialog format. The reader asks a question and then the developer answers it in the documentation.

Both people should work on the document at the same time, where only the reader speaks, and the writer communicates through the written word.

8.1.4. <u>Wiki</u>

Internal Wikis are essential tools in software development. Wikis allow us to create a pool of knowledge that everyone can agree upon.

This knowledge pool is essential for new team members, since it allows them to ramp up quickly.

It is also important for current team members, since it helps them clarify the requirements of the project.

An essential feature of wikis is that everyone must contribute, since everyone has knowledge to share.

 Comment: Wikis are always private and available only to the team responsible for the development of the product.

8.2. Engineering Documents

I cannot understand for the life of me why DOE is going forward with this licensing procedure when we do not know whether or not the scientific documentation upon which you are basing your decisions is, in fact, flawed.
-- Shelley Berkley --

Engineering documents are essential for all engineering endeavors. They define what we want to accomplish and the scope in which our solution will operate in. Without it, we don't have a unified vision of what we need to accomplish.

Keep in mind that everyone on the team is responsible for updating the documentation. The ideal way to store all documentation is as a WIKI, so everyone can maintain it.

8.2.1. Test Plan
An important architectural document is the test plan. This defines how the product will be tested and what scenarios are important.

The test plan has a few things in common.

Product Name
The title of the plan will be the title of the document. If the solution name is, Solution1, then the title would be: Soultion1 Test Document (Standard stuff).

Product History
Here we include the history of the product, including any relevant updates to the product.

Description
Here is where we describe the product. You should include:
- Why the solution was created.
- The need the solution is intended to address
- Who the users are

- The type of solution: Single user application, windows service, network service, web service
- Platforms: Phone based, browser based, desktop, server

We also need to include links to all relevant wikis, as well as the network location of this document.

Technologies
Technologies used will be described here. This includes the framework used.
- .NET Framework
- Java Runtime
- Browser based
- Phone based
- Native code

Technologies also include network requirements and servers used.

Services are also included here.

Automation Framework
Here we describe the automation framework we will be using to validate the functionality of the product.

Included here is also the development environments used. Some popular ones, in alphabetical order are:
- Eclipse
- Icsharpcode
- Net Beans
- Visual Studios

We will also describe the types of testing we need to do. They include:
- Performance testing
- Security testing
- Unit testing

Don't forget your source control requirements, such as Git and TFS.

Each of the above should be in its own document, and preferably in the form of a WIKI.

8.2.2. __Feature Plan__

All features included in the finished product need a feature plan. The plan describes the feature, how it will be tested and the context in which it operates.

Feature Name
The feature name should be descriptive, but not too verbose.

Description
This is a brief description of the feature we are developing and maintaining.

Keywords
Each feature must have keywords that describe it. This is essential for managing and organizing functionality.

Ideally the feature should be completely describable using keywords.

Technologies
The features we are developing require technologies to implement. However, not all features use the same technologies.

A database connection uses different technologies than a UI element.

These requirements must be stated.

8.2.3. __Feature Test Cases__

Test cases are fundamental parts of the feature documentation. They define the feature in full detail.

Feature Name
The name of the test case must be as concise as possible.

Feature Link
A link to the feature to be tested...

Test Keywords
Here will be the test categories the test case will be in.

Test Setup
This is where we describe how to setup the environment where the test will run. Ideally we will point to pre-existing setups in order to create a consistent test environment.

Scenario
Here we describe the steps needed to exercise the feature.

Verification
Verification involves making sure that the feature behaves as expected.

8.2.4. Keywords
Keywords are the metadata that describes and organizes all functionality of a solution. This allows us to query each piece of functionality and see if it is described in sufficient detail as to satisfy the needs of the customer.

All keywords need to be fully defined in a way that everyone agrees. As with other documents, this will be in the form of a WIKI.

8.3.   Working S.M.A.R.T.

A man must be
Big enough to admit his mistakes,
Smart enough to profit from them,
and strong enough to correct them.
-- John C. Maxwell –

> **Comment:** Computers are stupid (i.e. we need to spell everything out for them) and we are smart. However, when we work with computers, we need to act stupid in order to be **S.M.A.R.T.**

8.3.1. Specific
All requirements must be as specific as possible. Fortunately there are some **S.O.L.I.D.** guidelines to help us. (See. 152)

8.3.2. Measurable
Measurement comes in two parts.

First, we need to know what to expect. This is satisfied by the requirements documents, as well as feature documents.

Second, we need to exercise the functionality. This is achieved through either manual or automated testing. This will be covered in more detail in Requirements Verification (P. 158).

8.3.3. Attainable

Based on the current level of existing technology, is our goal attainable? We don't know and we can't know until the feature is fully defined.

Once we know what we want, then we look at the technologies available and also the technologies embedded in our product. We can't use a technology that is incompatible with our solution.

This involves cost analysis, which is outside this scope of this book.

8.3.4. Realistic

Just because something is attainable, doesn't mean it is cost effective to implement it at this stage of the game. We can build a super car, but will enough people buy it to justify the cost?

8.3.5. Timely

Can we deliver the feature within the timeframe specified? That is a million dollar question. The scrum methodology was invented to manage this issue.

8.4. S.O.L.I.D. Principles

And now it's solid
Solid as a rock
-- Ashford And Simpson-Solid As A Rock --

Managing feature development and implementation is a cross-cutting concern that affects all aspects of software development.

Every design decision we make affects how the product evolves. Therefore to build smart we need a **S.O.L.I.D.** foundation.

S.O.L.I.D. principles are a cross-cutting concern that affects everyone, and not just something developers need to think about. This is why we present it here.

These principles allow us to verify that we have code that is:
*   Easy to use
*   Simple to maintain
*   Reusable in various scenarios

- Flexible enough to adapt to future needs.

 Take Home: The **S.O.L.I.D.** principles allow us to adhere to the more basic K.I.S.S. (Keep it simple) principle.

8.4.1. Single Theme, Single Responsibility

All features need to have a singular purpose. They need to do one thing and one thing only. Having a feature perform multiple actions is an invitation for trouble. It hampers refactoring and makes it harder for the feature to be used correctly.

This is especially true of class definitions. By keeping to one theme, we simplify the maintenance of the feature. It also makes the feature easier for clients to use.

This is especially true when implementing a library of functionality.

 Comment: All classes must be built around a single theme. All methods must have a single responsibility.

 Take Home: Build class definitions around a **single** noun or archetype. Build methods around a **single** simple and easy to understood verb.

EX: Data Retrieval

Imagine you want to connect to a database and retrieve user data.

1. We create a user manager class that only worries about users. This class will not implement all functionality. Instead it will delegate responsibility for connecting to the database to another helper {Chain-Of-Responsibility}.
2. We expose functionality directly related to managing users.

We then we assemble the pieces from the building blocks we have.

EX: Undo-Redo

A great example of software that breaks the single responsibility rule is Microsoft Office. MS Office has a mechanism that allows us to undo or redo changes to the document we are working on.

 To undo a change, click on icon, or press <Ctrl-Z>

Undo <Ctrl-Z>

So far, so good: Now the interesting part comes.

Open an MS Office document (Ex. Word) and press <Enter>, then move the caret (Blinking line) up. When we do that, something interesting happens – The functionality of the redo button changes. It changes from Redo to Repeat, and the icon changes as well.

 You get a nasty surprise if you are not paying attention and just pressing the button multiple times. The button changes from to redo to repeat and strange text appears.

Repeat action <Ctrl-Y>

Another example is Microsoft Excel. MS Excel shares one undo list among all open spreadsheets. As a result, there is a serious danger of data loss when working on multiple Excel spreadsheets at the same time.

Now imagine this happens with shipped code. Of course you can document the double behavior. However, who reads documentation?

8.4.2. Open/closed

Using the single-theme, single-responsibility principle, we create a feature that can be used on multiple projects.

Now what?

Sooner or later we discover that the class we created lacks functionality needed to satisfy feature requirements.

We have two choices: We can add functionality to the class, or we can create a new class based on the existing class

Bad: Modifying an Existing class Definition

If we add functionality, then existing programs would probably work. But can you be sure? Program state could have been modified.

To be absolutely sure, we will need to test.

This is especially true if existing functionality needs to be modified.

Good: Extending a Type

By extending a class type, we can add new functionality while ensuring users of the old type isn't affected.

Also, by overriding the implementation of a feature we open the possibility of polymorphism.

 Take Home: All class definitions should be open to extension but closed to modification.

 Note: Some classes in .NET are closed to both extension and modification. An example is the String class.

The solution in this case is to use the C# feature called extension methods, also known as {Decorators}

8.4.3. Liskov Substitution Principle

Liskov Substitution Principle simply means that one type can be replaced by another type, as long as the basic contract is satisfied.

This related to polymorphism (see P. 72), where behavior changes based on the implementation details.

The important thing here is that the calling code gets exactly what it expects based on the established contract.

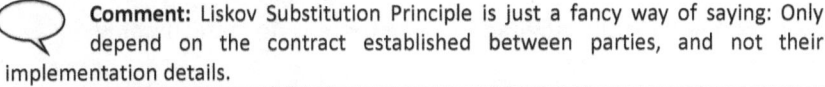 **Comment:** Liskov Substitution Principle is just a fancy way of saying: Only depend on the contract established between parties, and not their implementation details.

8.4.4. Interface segregation principle

In essence this means that we must create one contract (interface) for each scenario we will be dealing with. Each scenario must encapsulate a clearly defined theme.

This simplifies the interactions with clients and makes client code easier to debug.

The biggest challenge when dealing with this principle is that developers want to future proof the code. As a result, they add features they think they might need in the future.

The solution is to just solve the problem at hand. Then create new contracts to solve new scenarios. This ensures the design is always **S.O.L.I.D.**

 Take Home: By creating multiple contracts for each scenario or theme, we ensure that clients get exactly what they need. This also simplifies

development since users don't need to know about feature interactions (there is only one theme exposed per contract or interface).

8.4.5. Dependency inversion principle

"Inversion of Control (IoC): Inversion of Control (IoC) is an object-oriented programming practice whereby the object coupling is bound at runtime by an 'assembler' object and is typically not knowable at compile time using static analysis." – https://stackoverflow.com/users/3784758/nadhu

Dependency inversion is simple. We use it all the time when we use function pointers or delegates.

In the old days, we only worked with concrete implementations of code. We referenced a library and then called the methods on the library as needed. When compiled, the dependencies are locked in place.

In other words, all code dependencies are known and bound at application creation time.

With Inversion of Control, we don't know who or what will perform the operation. Instead we have a placeholder. At runtime the placeholder is replaced with a call to some library the developer has no knowledge of.

This is common in dynamic languages, since there is no static binding.

With static languages, it is common to use function pointers or delegates.

```
/// <summary>
/// We create a library that will be shipped to some third party.
/// </summary>
class Library1
{
    /// <summary>
    /// Since this is a shipped library,
    /// we don't know how unknownFunction is implimented.
    /// The only thing we know about unknownFunction is
    /// that it conforms to our contract.
    /// The dependency is only resolved at runtime
    /// </summary>
    /// <param name="unknownFunc"></param>
    public void PerformAction(Func<string, string> unknownFunc)
    {
        var returnedString = unknownFunc.Invoke("Some String");

        // Do something
    }
}
```

{Command} Design Pattern

The intent of the Inversion of Dependency pattern is to defer binding of an implementation until runtime.

The key to this is the Command design pattern. The Command encapsulates the call to a library method. This is then passed to the caller function as needed, when the function is called.

In C# this is implemented as the delegate. However its functionality is extended using IoC containers such as Unity.

In Windows' WPF, we have the ICommand interface. With this we define:
1. A way to poll if the command is ready.
2. An event that fires when we the command state changes between ready and not ready.
3. The actual command to execute.

```csharp
namespace System.Windows.Input
{
    public interface ICommand
    {
        event EventHandler CanExecuteChanged;
        bool CanExecute(object parameter);
        void Execute(object parameter);
    }
}
```

Inversion of Control Containers

There are many Inversion of Control tools we can use to defer implementation for later.

For .Net, the most famous IoC container is:
*   https://github.com/unitycontainer/unit

> **Comment:** Dependency Inversion allows us to add implementation details at runtime, through the use of dynamically bound libraries. As long as the library function honors the contract, then we are good to go.

8.5. Requirements Verification

Irreproducible bugs become highly reproducible
right after delivery to the customer
-- Michael Stahl's derivative of Murphy's Law --

Storytelling strikes me as a more powerful tool
than quantification or measurement
for what we do
-- Alan Cooper --

We now have a formalized set of requirements for a product we want to ship and we have the product we want to ship. How do we make sure that the product conforms to our requirements? The answer is requirements verification (also known as testing).

There are multiple forms of requirements verification: API testing, functional testing, load testing, globalization/localization testing, etc. Each form of verification addresses specific aspects of a solution and not all solutions require each type of verification.

Before we decide what kinds of verification we need, we need to understand our customers and their needs. This should be answered in our requirements document. If it is not, then the document needs to be updated.

8.6. Code Management

Without (code) management,
how can we hope to Ruuule the
(application) world?

In any programming endeavor we need to store and manage our code base. This is less important when there is only one developer. However, when dealing with multiple developers, code management becomes essential.

The biggest advantage with source control is allowing multiple people to work on the same file at the same time.

Another advantage of source control is being able to rollback changes when an impossible to find bug shows up while adding new functionality.

Imagine you need to refractor some code. You make a code change that breaks your code. Unfortunately you can't find the code break. So you rollback your changes, and then start over again.

8.6.1. <u>Source Control</u>

There are a number of source control solutions on the market. The most popular are Git and TFS.

<u>Git</u>

Git is an open source solution that anyone can install on their own private network.

GitHub is the service for hosting public projects for free. It can also host private repositories for a yearly fee.

<u>Team Foundation Server (TFS)</u>

Microsoft's solution to source control is Team Foundation Server (TFS).

As with Git, you can host your own server on your own equipment. However, it is a paid product.

 Comment: For those with teams of 5 or less people TFS is free. For details, See: https://azure.microsoft.com/en-us/solutions/devops/

8.6.2. <u>Refactoring</u>

Code that can't be freely re-factored is code that isn't worth keeping and will cost you your trip to Hawaii.

An interesting thing about refactoring code is that it reveals flaws in your test automation, as well as hidden assumptions about the product that may or may not be valid. This allows us to formally track these assumptions, making the code more robust.

Refactoring is also a great way to understand the code base and the requirements.

 Comment: One of the interesting things about refactoring code is that it reveals hidden assumptions about our requirements.

8.6.3. <u>Code Review</u>

Our code base is constantly being updated. This is needed to add functionality and fix bugs.

However, adding or modifying code is a risk.

- The code many be hard to read
- The change may not use the best algorithm

- The code may be duplicating functionality
- The code may not conform to the standards used in your company.

As a result, it is necessary for your code to be reviewed by a knowledgeable person in your group.

When doing code reviews, keep these points in mind:

Be available

It is essential that the code reviewer be available to answer questions regarding the code review. The reason is that, sometimes a developer makes changes for reasons the reviewer doesn't understand. If the reviewer is not available for discussion, the developer might be pressured into including the suggested changes and submitting the code. This is done in spite of the possibility that the suggestions given by the reviewer might be invalid.

Beware of the Code Czar

Keep in mind, too much control over source code will result in poor design. This is especially true when you don't have full buy in.

As an example, I was on a team that had a code Czar. His job was to ensure the quality of the code base. My manager was scared of him and was always trying to sneak code changes behind his back. She instructed me on ways to do the same and told me to be quiet when he was in the room.

Such secretive behavior only hurts the product and ultimately the customer and the bottom line.

Unified Vision

The above reveals an important lesson. Code quality is impossible unless you have complete buy in from everyone.

It is essential that everyone understands the rules the team follows. It is even more important for the team be open to changing the rules when necessary.

9. APPENDIX

9.1. Keyboard Shortcuts

Here are some of the most common keyboard commands I have encountered, in Microsoft Word and other applications.

CTRL-A	Select All		CTRL-N	New document
CTRL-B	Bold		CTRL-O	Open file dialog
CTRL-C	Copy		CTRL-P	Paste
CTRL-D			CTRL-Q	
CTRL-E			CTRL-R	
CTRL-F	Find dialog		CTRL-S	Save document
CTRL-G	Go to document location		CTRL-T	
CTRL-H	Replace text		CTRL-U	
CTRL-I	Italics		CTRL-V	
CTRL-J			CTRL-W	
CTRL-K	Insert Hyperlink		CTRL-X	Cut selected text
CTRL-L	Left Justify		CTRL-Y	Redo action
CTRL-M			CTRL-Z	Undo action

CTRL-Page-Down	Find next		Home	Beginning of line
CTRL-Page-Up	Find previous		End	End of line
CTRL-Home	Beginning of document			
CTRL-End	End of document			

F1	Help		F7	
F2			F8	
F3	Find Next		F9	
F4			F10	
F5			F11	
F6			F12	

9.2. **Order of an Operation O(n)**

Judge: Order in the Court
Defendant: Your Honor I'll have a pint
-- Unknown --

Finding out how long an operation takes place is essential to making sure that we have a good user experience.

If the algorithm is working on a single item, then the time it takes will be approximately constant. This assumes that any external calls are also operated in constant time.

9.2.1. **Data Collections**

For collections, the time it takes to perform an action can vary greatly. As a general rule, we can find the order of an operation by counting the number of embedded loops used.

Constant time
For some collections types we can find our data immediately. The perfect example is the array. Given an index, you go to data in one step.

Linear time
If you need to go through a collection once, then the order of operation is linearly proportional to the length of the collection.

Log time
Some algorithms, such as binary search, can find data much faster than going from one element to another.

Exponential Time
With exponential time, the time it takes to execute an algorithm increases exponentially with the number of elements in the collection.

To find the time:
9.   Count the number of embedded loops. Assign it to n.
10. The count would then be $Count^n$.
11. If multiple loops exist, then the one with the greatest number of embedded loops determine the order of the operation.

In the below example, there are two embedded loops. Therefore the order of operation is $O(n^2)$.

```
/// <summary>
/// Simple bubble sort. O(n2).
```

```
/// </summary>
void BubbleSort(int[] arr)
{
        if (arr == null || arr.Length <= 1) return;

        // Loop 1
        for (int i = 0; i < arr.Length - 1; i++)
        {
                // Inner loop 2
                for (int j = i + 1; j < arr.Length; j++)
                {
                        if (arr[i] > arr[j])
                        {
                                arr[i] ^= arr[j];
                                arr[j] ^= arr[i];
                                arr[i] ^= arr[j];
                        }
                }
        }
}
```

9.3. Pass-Made-Rugs

Fashion is a language that creates itself in clothes to interpret reality.
-- Karl Lagerfeld --

Everything is design. Everything!
-- Paul Rand --

1. **P**erformance
    1. **P**erformance Counters.
    2. **A**lgorithm.
    3. **C**ode Tweaking.
    4. **E**nvironment
2. **A**vailability
    1. **H**ardware
    2. **A**vailable services
    3. **N**etwork issues
    4. **D**istributed platforms
3. **S**ecurity
    1. **S**poofing
    2. **T**ampering
    3. **R**epudiation
    4. **I**nformation Disclosure
    5. **D**enial of Services
    6. **E**levation of Privileges
4. **S**calability
    1. **M**ultiple servers

       2. **A**pplication pooling
       3. **D**ata warehouses
5. **M**aintainability
       1. **D**ocumentation
       2. **O**bject-oriented programming
       3. **L**ogging
       4. **T**est automation
6. **A**ccessibility
       1. **H**earing
       2. **I**mpaired vision
       3. **P**hones
7. **D**eplorability
       1. Web Applications
       2. Platform Applications
       3. Installers
       4. Single-User Applications
       5. Multi-Server Solution
       6. Cloud-Based Solutions
       7. User Installation
       8. Bug-Fixes, Service/Feature Packs
       9. Repair Installation
       10. Clean Uninstall
8. **E**xtensibility
       1. Plug-in model
9. **R**esponsiveness
       1. **B**usy Indicators
       2. **U**I Responsiveness
       3. **R**esource Management
       4. **P**rogram Startup
10. **U**sability
       1. **D**esign Elements
       2. **A**nnoying Popups
       3. **D**ocumentation
       4. **S**DKs
11. **G**lobalization
       1. **L**ocal Units
       2. **I**nformation Text
       3. **D**aylight Savings Time
       4. **S**ound and Video
12. **S**ex Appeal

9.4.   **Password Security**

Security is always excessive until it's not enough.
-- Robbie Sinclair, Head of Security, Country Energy, NSW Australia --

Passwords are the standard way of securing user accounts. As such, their management is vital to system security. Here are some things to keep in mind while managing passwords.

Password Hiding

Password hiding is common in the Unix/Linux world. With this method, the user enters the password, but nothing is displayed.

The advantage is that no one can know the length of the password. The disadvantage is that it makes it difficult for the user to know if they entered the correct password.

The standard now is to display a dot • for every character entered. That way, they can be certain when they delete characters from the password.

Blank password

Sometimes I accidently press {Enter} when typing the user name. As a result, a password of zero length is sent.

Zero-length passwords should not be considered as an incorrect password, since it is impossible to have a zero-length password in a properly configures system.

Password Retries

Many times a user mistypes a password. That is only natural. In my opinion, a user should be allowed to enter an incorrect password at least 3-5 times, provided there is a time delay of at least one second between retries.

Same Incorrect Password

Sometimes I enter a password I'm convinced is correct and it is rejected as being incorrect. As a result I reenter the same password. I then realize that I was using the wrong password, but it's too late.

As a courtesy to the user, you might consider marking multiple entries of the same password as a single try, provided they are consecutive.

Example of three tries:
1.   WrongPassword1
2.   WrongPassword1
3.   WrongPassword1

4. WrongPassword2
5. CorrectPassword1

Example of four tries:
1. WrongPassword1
2. WrongPassword2
3. WrongPassword1
4. WrongPassword1
5. CorrectPassword1

Incorrect Password Response
When creating a user account, it is essential to have a way of contacting the user to notify them when you believe the account has been compromised. This will prevent user aggravation. No one likes a locked account.

Consider sending a link to the legitimate user that allows the account to be unlocked. The alternate would be for the user to make a phone call to an automated customer service system.

Password Peaking
Some systems allow the user a peek at the passwords they are typing, to ensure that the password is correct.

A notable example of this is Internet Explorer. Clicking on the eye reveals the password. However this functionality is only available while you are typing. Shifting the focus away from the password box and bringing it back disables the functionality. It will only be enabled when you delete the password entirely.

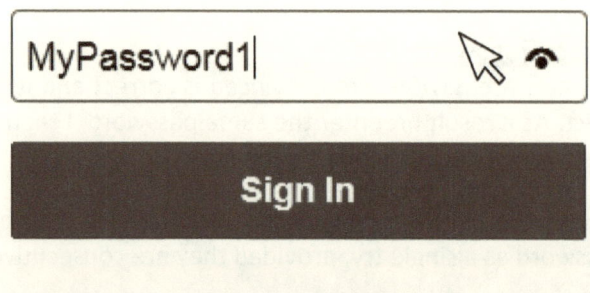

Screen shot of password text box in Internet Explorer.

On cell phones, the character you type is visible for a second. This is necessary because of the nature of cell phones. It is very easy to mistype on a cell phone, making it difficult to sign onto services using your phone.

Pasting Passwords

Many times a security system allows the user to paste the password. Many times this is unintended on the part of the system designer, since they are using standard UI elements.

I like it when it is allowed since it is convenient for me. It ensures I don't make a mistake.

Some systems try to prevent this, supposedly for security reasons.

This serves no practical purpose, and is just an annoyance to the user. If the bad guy has the password, then the system can't stop the user from manually typing the correct password.

It is true that a script could automatically try multiple passwords. However, the system should be designed to detect when multiple incorrect passwords are typed consecutively. In this respect, it is no different from manually typing in multiple consecutive incorrect passwords.

Copying Passwords

The copying of passwords is never allowed in normal systems.

Passwords Expiration

To counter stolen passwords, many companies require users to periodically change their login password. This renders stolen passwords useless, but can be a pain for users.

Session Timeout

One way bad people get illegal entry is when a legitimate user logs on and walks away for whatever reason. When this happens, anyone can come and hijack the session. The legitimate user would then be responsible should the hacker do something bad.

The most common countermeasure is the timeout. If the user doesn't interact with the service for a certain period of time, the service logs them out.

Timeouts are common on network connected computers. The screensaver comes on after a certain timeout (Say five minutes of inactivity), requiring the user needs to re-login to resume working.

Note that timeouts on networked computers can be a pain in the ass and be countered by either software or hardware mouse jigglers.

A software mouse jiggler is a program that moves the mouse pointer by a tiny amount every second or so. The counter for this is preventing software from being installed.

For those working for paranoid companies, another solution is the hardware mouse jiggler is. The hardware mouse jiggler is a USB device that the system recognizes as a USB mouse.

The counter for the hardware mouse jiggler is making sure the user only has one mouse plugged into the computer. For those working for such companies, consider this: Is it worth working for someone so paranoid?

Challenge Question

To guard against hackers, companies use security questions which only the company and the person should know. These questions are asked when the account is created.

Where were you born?	Land of Cools-Ville
What was your first pet's name	Snoopy

Verification Codes

In addition, companies us verification codes that presumably computers can't crack. Anyone who created a new email address has seen this:

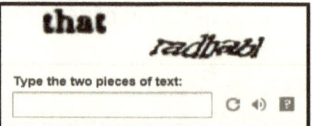

The user types in the challenge text into the text box. Once this is done, an account is created or enabled.

This tool is designed to foiling computers, but it can foil users as well. What is that word on the right?

 Warning: Machine learning is already capable of overcoming the code verification test.

Anti-Automation Controls

Many hackers use automation tools to spam web sites with fraudulent requests. This is a problem that affects all users of the system.

To protect their services, many sites use this type of control to verify that a human is submitting the data. Of course, this relies on support from the web browser.

Also, the service provider can't rely on this. A determined enough hacker can easily bypass this.

9.5. Useful classes

Planning ahead is a measure of class. The rich and even
the middle class plan for future generations, but the poor
can plan ahead only a few weeks or days.
-- Gloria Steinem --

You've got to have class, loads and loads and loads of class.

9.5.1. Data Mode Base Class

This is an implementation of the data model base class I use to manage the interaction between the data model and the view used to display the model to the user.

This was originally created for use with Microsoft's Universal Windows Platform (UWP).

The key element is that this eliminates the need to have class-level variables for storing model state.

This is used in the solution: github.com/TrevyBurgess/ZeeEventsManager

```
//------------------------------------------------------------
// <copyright file="ViewModelBase.cs" company="CyberFeedForward" >
// Free for use, modification and distribution
// </copyright>
// <Author>
// Trevy Burgess
// </Author>
//------------------------------------------------------------
namespace CyberFeedForward.WUP.Common.WPF
{
    using System;
    using System.Collections.Generic;
    using System.ComponentModel;
    using System.Runtime.CompilerServices;
    using Windows.Storage;

    /// <summary>
    /// Implementation of <see cref="INotifyPropertyChanged"/> to simplify
models.
    /// For Data Capacity: https://msdn.microsoft.com/en-
```

```
us/library/windows/apps/windows.storage.applicationdata.roamingsettings.aspx
    /// </summary>
    public abstract class ViewModelBase : INotifyPropertyChanged
    {
        protected ViewModelBase()
        {
            propertyStore = new Dictionary<string, object>();
        }

        /// <summary>
        /// Backing store for properties
        /// </summary>
        private Dictionary<string, object> propertyStore;

        /// <summary>
        /// Multicast event for property change notifications.
        /// </summary>
        public event PropertyChangedEventHandler PropertyChanged;

        /// <summary>
        /// For use with calculated properties.
        /// Notify listeners that the specified property has changed. Only
use in properties.
        /// </summary>
        /// <param name="calculatedPropertyName">Calculated property
name</param>
        protected void NotifyPropertyUpdated(string calculatedPropertyName)
        {
            OnStateChanged(calculatedPropertyName);
        }

        /// <summary>
        /// Set property state.
        /// </summary>
        /// <typeparam name="T">Type of the property</typeparam>
        /// <param name="value">Desired value for the property</param>
        /// <param name="propertyName">Do not set</param>
        /// <returns>True if the value was changed, false otherwise</returns>
        protected bool SetState<T>(
            T value,
            SaveType saveType = SaveType.Application,
            [CallerMemberName] string propertyName = null)
        {
            switch (saveType)
            {
                case SaveType.Application:
                    if (propertyStore.ContainsKey(propertyName))
                    {
                        if (Equals(propertyStore[propertyName], value))
                        {
                            return false;
                        }
                        else
                        {
                            propertyStore[propertyName] = value;
                            OnStateChanged(propertyName);
                            return true;
                        }
                    }
                    else
```

```
                                {
                                    propertyStore[propertyName] = value;
                                    return true;
                                }

                        case SaveType.RoamingSettings:
                            if
(ApplicationData.Current.RoamingSettings.Values[propertyName] == null)
                                {

ApplicationData.Current.RoamingSettings.Values[propertyName] = value;
                                    ApplicationData.Current.SignalDataChanged();
                                    OnStateChanged(propertyName);
                                    return true;
                                }
                            else if
(Equals(ApplicationData.Current.RoamingSettings.Values[propertyName], value))
                                {
                                    return false;
                                }
                            else
                                {

ApplicationData.Current.RoamingSettings.Values[propertyName] = value;
                                    ApplicationData.Current.SignalDataChanged();
                                    OnStateChanged(propertyName);
                                    return true;
                                }

                        case SaveType.LocalSettings:
                            if
(ApplicationData.Current.LocalSettings.Values[propertyName] == null)
                                {

ApplicationData.Current.LocalSettings.Values[propertyName] = value;
                                    ApplicationData.Current.SignalDataChanged();
                                    OnStateChanged(propertyName);
                                    return true;
                                }
                            else if
(Equals(ApplicationData.Current.LocalSettings.Values[propertyName], value))
                                {
                                    return false;
                                }
                            else
                                {

ApplicationData.Current.LocalSettings.Values[propertyName] = value;
                                    ApplicationData.Current.SignalDataChanged();
                                    OnStateChanged(propertyName);
                                    return true;
                                }

                        default:
                            throw new NotImplementedException();
                    }
                }

            /// <summary>
            /// Retrieve stored data
```

```
        /// </summary>
        /// <typeparam name="T">Data type to store</typeparam>
        /// <param name="initialValue">Initial value to set</param>
        /// <param name="saveType">Save type</param>
        /// <param name="propertyName">Leave blank.</param>
        /// <returns>Get stored value</returns>
        protected T GetState<T>(
        T initialValue = default(T),
            SaveType saveType = SaveType.Application,
            [CallerMemberName] string propertyName = null)
        {
            switch (saveType)
            {
                case SaveType.Application:
                    if (!propertyStore.ContainsKey(propertyName))
                    {
                        propertyStore[propertyName] = initialValue;
                    }

                    return (T)propertyStore[propertyName];

                case SaveType.RoamingSettings:
                    if
(ApplicationData.Current.RoamingSettings.Values[propertyName] == null)
                        {

ApplicationData.Current.RoamingSettings.Values[propertyName] = initialValue;
                        ApplicationData.Current.SignalDataChanged();
                        }

                    return
(T)ApplicationData.Current.RoamingSettings.Values[propertyName];

                case SaveType.LocalSettings:
                    if
(ApplicationData.Current.LocalSettings.Values[propertyName] == null)
                        {

ApplicationData.Current.LocalSettings.Values[propertyName] = initialValue;
                        ApplicationData.Current.SignalDataChanged();
                        }

                    return
(T)ApplicationData.Current.LocalSettings.Values[propertyName];

                default:
                    throw new NotImplementedException();
            }
        }

        /// <summary>
        /// Get relay command for the specified action. Only use in get
property.
        /// </summary>
        /// <param name="execute">The action</param>
        /// <param name="canExecute">Is enabled Function</param>
        /// <param name="propertyName">Property name</param>
        /// <returns>Relay command</returns>
        protected RelayCommand Command(Action execute, Func<bool> canExecute
= null, [CallerMemberName] string propertyName = null)
```

```csharp
            {
                if (!propertyStore.ContainsKey(propertyName))
                {
                    propertyStore[propertyName] = new RelayCommand(execute,
canExecute);
                }

                return propertyStore[propertyName] as RelayCommand;
            }

            /// <summary>
            /// Get relay command for the specified action. Only use in get
property
            /// </summary>
            /// <typeparam name="TCommandParameter">WPF
<code>CommandParameter</code></typeparam>
            /// <param name="execute">The action</param>
            /// <param name="canExecute">Is enabled Function</param>
            /// <param name="propertyName">Do not set</param>
            /// <returns>Relay command</returns>
            protected RelayCommand<TCommandParameter> Command<TCommandParameter>(
                Action<TCommandParameter> execute,
                Func<TCommandParameter, bool> canExecute = null,
                [CallerMemberName] string propertyName = null)
            {
                if (!propertyStore.ContainsKey(propertyName))
                {
                    propertyStore[propertyName] = new
RelayCommand<TCommandParameter>(execute, canExecute);
                }

                return
(RelayCommand<TCommandParameter>)propertyStore[propertyName];
            }

            /// <summary>
            /// Notifies listeners that a property value has changed
            /// </summary>
            /// <param name="propertyName">Property whose name has been
changed</param>
            private void OnStateChanged(string propertyName)
            {
                PropertyChanged?.Invoke(this, new
PropertyChangedEventArgs(propertyName));
            }
    }

    /// <summary>
    /// View model Storage type.
    /// </summary>
    public enum SaveType
    {
        /// <summary>
        /// Application storage storage for settings
        /// </summary>
        Application,

        /// <summary>
        /// Local persistant storage for settings
        /// </summary>
```

```
            LocalSettings,

            /// <summary>
            /// Roaming persistant storage for settings
            /// </summary>
            RoamingSettings
    }
}
```

9.5.2. Relay Command Class

The data model base class defined above relies on the next two classes...

```
//----------------------------------------------------------
// <copyright file="RelayCommand.cs" company="CyberFeedForward" >
// Free for use, modification and distribution
// </copyright>
// <Author>
// Trevy Burgess
// </Author>
//----------------------------------------------------------
namespace CyberFeedForward.WUP.Common.WPF
{
    using System;
    using System.Windows.Input;

    /// <summary>
    /// Implementation of ICommand for use with MVVM WPF applications.
    /// </summary>
    public class RelayCommand : ICommand
    {
        /// <summary>
        /// Command to execute
        /// </summary>
        private readonly Action execute;

        /// <summary>
        /// Return true if execute action available, false otherwise
        /// </summary>
        private readonly Func<bool> canExecute;

        /// <summary>
        /// Initializes a new instance of the <see cref="RelayCommand" />
class.
        /// </summary>
        /// <param name="execute">The execution logic.</param>
        /// <param name="canExecute">The execution status logic.</param>
        public RelayCommand(Action execute, Func<bool> canExecute = null)
        {
            if (execute == null)
            {
                throw new ArgumentNullException("execute");
            }

            this.execute = execute;
            this.canExecute = canExecute;
        }

        /// <summary>
```

```
            /// Raised when RaiseCanExecuteChanged is called.
            /// </summary>
            public event EventHandler CanExecuteChanged;

            /// <summary>
            /// Determines whether this <see cref="RelayCommand"/> can execute in
its current state.
            /// </summary>
            /// <param name="parameter">Data used by the command.</param>
            /// <returns>True if this command can be executed, false
otherwise.</returns>
            public bool CanExecute(object parameter = null)
            {
                return canExecute == null ? true : canExecute();
            }

            /// <summary>
            /// Executes the <see cref="RelayCommand"/> on the current command
target.
            /// </summary>
            /// <param name="parameter">Data used by the command.</param>
            public void Execute(object parameter = null)
            {
                execute();
            }

            /// <summary>
            /// Method used to raise the <see cref="CanExecuteChanged"/> event
            /// to indicate that the return value of the <see cref="CanExecute"/>
            /// method has changed.
            /// </summary>
            public void RaiseCanExecuteChanged()
            {
                CanExecuteChanged?.Invoke(this, EventArgs.Empty);
            }
        }
}
```

9.5.3. <u>Relay Command <T> Class</u>

This is the generic version of the relay command.

```
//-------------------------------------------------------------
// <copyright file="RelayCommand.cs" company="CyberFeedForward" >
// Free for use, modification and distribution
// </copyright>
// <Author>
// Trevy Burgess
// </Author>
//-------------------------------------------------------------
[module: System.Diagnostics.CodeAnalysis.SuppressMessage(
    "StyleCop.CSharp.DocumentationRules",
    "SA1638:FileHeaderFileNameDocumentationMustMatchFileName",
    Justification = "Bug in Style Cop.")]

namespace CyberFeedForward.WUP.Common.WPF
{
    using System;
    using System.Windows.Input;
```

```
    /// <summary>
    /// Implementation of ICommand for use with MVVM WPF applications.
    /// </summary>
    /// <typeparam name="TCommandParameter">WPF argument</typeparam>
    public class RelayCommand<TCommandParameter> : ICommand
    {
        /// <summary>
        /// Command to execute
        /// </summary>
        /// <typeparam name="TCommandParameter">Action parameter</typeparam>
        private readonly Action<TCommandParameter> execute;

        /// <summary>
        /// Return true if execute action available, false otherwise
        /// </summary>
        /// <typeparam name="TCommandParameter">Action parameter</typeparam>
        /// <typeparam name="bool">return parameter</typeparam>
        private readonly Func<TCommandParameter, bool> canExecute;

        /// <summary>
        /// Initializes a new instance of the <see
cref="RelayCommand{TCommandParameter}" /> class.
        /// </summary>
        /// <param name="execute">The execution logic.</param>
        /// <param name="canExecute">The execution status logic.</param>
        public RelayCommand(Action<TCommandParameter> execute,
Func<TCommandParameter, bool> canExecute = null)
        {
            if (execute == null)
            {
                throw new ArgumentNullException("execute");
            }

            this.execute = execute;
            this.canExecute = canExecute;
        }


        /// <summary>
        /// Raised when RaiseCanExecuteChanged is called.
        /// </summary>
        public event EventHandler CanExecuteChanged;

        /// <summary>
        /// Determines whether this <see cref="RelayCommand"/> can execute in
its current state.
        /// </summary>
        /// <param name="parameter">Data used by the command.</param>
        public void Execute(object parameter = null)
        {
            execute((TCommandParameter)parameter);
        }


        /// <summary>
        /// Called by WPF. Executes the <see cref="RelayCommand"/> on the
current command target.
        /// </summary>
        /// <param name="parameter">Data used by the command</param>
        /// <returns>True if control should be enabled</returns>
        public bool CanExecute(object parameter = null)
```

```
    {
        return canExecute == null ? true :
canExecute((TCommandParameter)parameter);
    }

        /// <summary>
        /// Method used to raise the <see cref="CanExecuteChanged"/> event
        /// to indicate that the return value of the <see cref="CanExecute"/>
        /// method has changed.
        /// </summary>
        public void RaiseCanExecuteChanged()
        {
            CanExecuteChanged?.Invoke(this, EventArgs.Empty);
        }
    }
}
```

9.6. Conducting Technical Interviews

To do a really good interview,
you have to be truly interested in the person.
-- Daisy Fuentes --

Here are a few thoughts on conducting technical interviews.

I went for an interview with a high-tech company. They were looking for programmers with several years of programming experience. I was expecting them to ask questions about my programming experience.

Instead, they asked questions that resembled a 1st year computer class exam:

- What is an interface?
- What is a class?
- What is polymorphism?

These questions don't reveal anything about a person's experience or abilities. In fact, these questions are worse than useless.

These questions penalize experienced developers. After years of industry experience, the answers to these questions are forgotten. I may not be able to formally define an interface, but I've used interfaces in countless situations. Ask me a question regarding a real-world problem and I will be able to answer it with ease.

On the other hand, a newbie will be able to tell you what an interface is, but may not know how to use it in real-world situations.

9.6.1. <u>The 30 Second Interview</u>

Countless studies have shown that we decide whether we like a person in the first 30 seconds. The rest of the interview is used to justify our feelings.

An interview is an unnatural situation to many people. It causes them to shut down and not act normally. This means that we are not looking at people at their best. Countless excellent candidates are passed over because of this. On the other hand, some people act confidently even though their abilities don't justify it. A good interviewer needs to take this into consideration.

9.6.2. <u>So how do we select someone to hire?</u>

Past competence is a good indication of future worth. If a person does well in the past, it's a good indication that they will do well in the future. The above company should have asked what programming challenges I faced and overcame. This would allow me to relate my experiences in terms of their needs.

The alternate is not necessarily true. Just because they did badly in the past, doesn't mean they will do badly now.

9.6.3. <u>So how do we select the best candidate?</u>

In my opinion, the only way to know the qualifications of a person is to work with them for at least a month. You put a new hire on probation without obligations on both sides. After the probation period ends, you have the choice of letting the contract end, or offering the person a permanent position in the organization.

9.6.4. <u>So how do we select someone for this probation period?</u>

For candidates, the best companies turn to their network. Word of mouth is the No 1 method of finding people. Companies look to their network to find out about prospects.

In the Brave New World we have entered, there is a new way of finding about candidates. This is the Internet. This is especially true of the social networking sites such as LinkedIn, FaceBook, and Twitter, to name but three.

Of course, this won't replace face-2-face interviews. I think the interview is like a date. Each is trying to decide if they can live with the other person. The technical interview is just a means to that end.

Microsoft Technical Interview Questions

Microsoft likes giving programming questions in their technical interviews. These questions are useful in finding out how people think. As a result, it doesn't matter if the candidate gets it right. The important thing is that the candidate explains their logic.

It's fundamentally important that the interviewer knows how to solve the programming question before they give it to the candidate. There is nothing worse than questioning a candidate and then having the candidate correct your mistakes. It's embarrassing for both you and your company. It's also bad for the candidate because now the interviewer resents them.

I was in this situation myself. I felt that the interviewer was incompetent and wondered why he got hired. Needless to say I didn't get the job.

Contract to Hire

Many companies hire out contractors to work for them instead of hiring employees. The company goes to a contract firm and gives them a list of requirements. The firm then sends candidates the company interviews.

The important thing about the contract is that it can be broken at any time without reason or penalty.

As a result, you have the option of working with a candidate short-term and seeing if they are worth keeping.

Home Grown Diamonds

Many companies try to steal the best people from other companies. In my opinion, this is a lose-lose situation.

We all know that it harms the company that is losing talent. However it doesn't stop there.

It also harms the company that is stealing the talent. They have to pay a premium for this talent. Also, what's to stop someone else from stealing this person? In my opinion, this is why executives get overpaid.

The best solution is to take a good worker and turn them into a Star.

A good candidate has these qualities:
1. Easy to work with.
2. Intelligent.
3. Dedicated to self-improvement.
4. Experienced in the specific technologies your business depends on.

Find a candidate with these qualities, and then help them to be a Star at work. To this end, I would like to suggest to you my favorite book on the subject of productivity. It's called, "How to Be a Star at Work: 9 Breakthrough Strategies You Need to Succeed"

9.7. Technical Interview Questions
The reporter claimed he was going to write the article from my point of view.
Instead, he made me sound like a little idiot.
It made me never want to do another interview again.
-- Tia Carrere --

Here are some technical questions I have collected over the years. It's best you work out the answers to these questions yourself. If you cheat, you will only hurt your company and yourself. I'm only including questions I myself have solved. I will NOT give out any answers because I think that's a disservice to the community.

In each category, the questions are arranged from simple to advanced.

9.7.1. Arrays 1-D
- Reverse an array of integers.
- Shuffle Cards.
- Bubble Sort an array of integers.
- Find element in an array of integers using binary search.
- Remove duplicates from sorted array of integers. Pad remaining cells with zeros
- Remove duplicates from unsorted array of integers. Pad remaining cells with zeros
- Find first element in array that is duplicated/not duplicated.
- Given an array of size N in which every number is between 1 and N, determine if there are any duplicates in it.
- Given an array t[100] which contains numbers between 1..99. Return the duplicated value. Use only one iteration, one int to store intermediate results.

- Given an array containing both positive and negative integers, find the sub-array with the largest sum. What if you don't want negative numbers in the returned sub-array?
- Implement the following function, FindSortedArrayRotation(int [] arr). arr is an array of unique integers that has been sorted in ascending order, then rotated by an unknown amount X, where 0 <= X <= (arrayLength - 1). An array rotation by amount X moves every element array[i] to array[(i + X) % arrayLength]. Can it be done in less than linear time?

9.7.2. Arrays 2-D (Matrix)
o Transpose a square matrix. I.e. interchange the rows and columns.
o Given an int array, zero out the vertical and horizontal column containing a zero.
o Find saddle point. A matrix is said to have a saddle point if entry arr[r, c] is the smallest value in the ith row and the largest value in the jth column. A matrix may have more than one saddle point.
o Find plateau. A matrix is said to have a plateau point if entry arr[r, c] is the largest value in both the rth row and the cth column. A matrix may have more than one saddle point.
- Write a routine that prints out a 2-D array in spiral order!

9.7.3. Binary Search Tree
- Insert node into binary search tree.
- Delete node from binary search tree.
o Write a function to find the depth of a binary tree.
o Find Lowest Common Ancestor, given: Node LCA(Node root, int left, int right);
o Print elements in-order (L, C, R).
- Convert BST into a linked list.

9.7.4. Linked Lists
- Insert element into linked list.
- Delete element from linked list.
- Delete alternate nodes in doubly linked list.
o Find middle element.
o Merge two sorted lists.
o Remove duplicates in sorted list.
o Find if list is circular. (No beginning or end node)
- Find M^{th} to last element in linked list.

- Reverse linked list.
- Delete the nth node, given a link to that node. You have no access to the previous node for the exercise.

9.7.5. Queues & Stacks
- Implement a queue only with stacks.

9.7.6. Text Strings
- Given a string s, find character c and return its position.
- Write a function to check if a given string is a palindrome. Ignore white spaces. Ex. "Able was I ere I saw Elba"
- Given an arbitrary string, return a string with no duplicate characters. String RemDup(String s)
- Given two strings S1 and S2, remove characters in S2 that appear in S1.
- Given two strings, write a function that would print characters in the first string that are not in the second string and also print characters in the second string that are not in the first string.
- Reverse words in String. (I am Sam) => (Sam am I)

9.7.7. Bit Operations
- Multiply num by 7 without using multiplication (*) operator.
- Find number of set bits in an integer.
- There are two integers, 'a' and 'b'. Swap the contents without using a temp variable.
- Print integer in base 2.General
- Find nth factorial. (0, 1, 2, 3) => (1, 1, 2, 6)
- Find nth Fibonacci number. (1, 2, 3, 4, 5, 6) => (1, 1, 2, 3, 5, 8)
- Print integer one digit at a time, without storing any intermediate results.
- Given the time in hours an minutes, find the angle between the hour hand and the minute hand.
- Write a function that takes as its arguments 3 integers > 0. If the product of the integers is even, then return the one with the lowest value, else show 0.
- Parse integer from string.

9.7.8. Recursion
- Print numbers from 1 to 100 and then 100 to 1 without any for or while loops.
- Print linked list in reverse order.

- o Reverse a singly linked list recursively.
- o Find nth factorial. (0, 1, 2, 3) => (1, 1, 2, 6)
- o Find nth Fibonacci number. (0, 1, 2, 3) => (1, 1, 2, 6)
- o Sum the digits in a number. Ex: If x is 2345, return 9 [2 + 3 + 4 + 5])
- ▪ Count the number of set bits in an array of integers without loops.
- ▪ Print all permutations in a string.
- ▪ Count square clusters in grid.
- ▪ Find way out of maze.

10. GLOSSARY

When a true genius appears in the world, you may know him by this sign -
that the dunces are all in confederacy against him.
-- Jonathan Swift --

The beginning of knowledge is the discovery of something we do not understand.
-- Frank Herbert --

Abstract Factory (Design Pattern): Abstract factories are similar to factories. How they implement their functionality by leveraging the power of multiple factories internally to fulfill their contract responsibilities.

Agent: An agent is a reference type object that acts on data objects (flyweights) and responds to commands.

Attribute: .NET metadata used to describe solution features, such as assemblies, classes, structs, enumerations, etc. They are more descriptive than tags, since they have functionality associated with them.

Bridge (Design Pattern): The separation of an abstraction (method definition) from its implementations (method implementation). It expresses itself whenever we create classes that extend an interface.

Class: The language construct used to define the properties and methods of a reference type object.

Class Definition: A class definition represents a singular noun in the business space of the solution. It defines the features of the noun (methods) and its capabilities (functions).

Data Objects: *See* Flyweight.

Chain of Responsibility (Design Pattern): An agent passes on the responsibility of fulfilling a request when it is unable to process the request.

Command (Design Pattern): An object that encapsulates a command to an agent.

Contract: A formally defined agreement established by multiple agents. This is expressed in a programming construct called an interface.

Design Pattern: A method for communicating design principles between members of the feature development world.

Encapsulation: The first pillar of object Oriented Programming. It allows developers to hide implementation details, allowing for a decoupling of implement and consumption.

Enumeration (Enum): A named list of numbers, used to define a finite set of choices a user or agent can select from. Some languages allow for a named list of prototypes.

Façade (Design Pattern): The intent of the façade pattern is to simplify our interaction with a complicated system. Each façade exposes only the features required by the agent or user in question. (*See* Interface)

Factory (Design Pattern): A factory is an encapsulated strategy for selecting an appropriate implementation that satisfies a required contract.

Flyweight (Design Pattern): An object that encapsulates an individual piece of business data. This is sharable with multiple agents and is usually immutable.

Immutable: An object that can't change its state once it is created.

Interface: A contract that formalizes the way the system interacts with agents and users.

Inversion of Control (Design Pattern): Dependency Inversion allows us to add implementation details at runtime, through the use of dynamically bound libraries.

Iterator (Design Pattern): A language construct that allows you to sequentially access all the elements of a data collection, in a forward-only way.

Mediator (Design Pattern): A {Mediator} is any process that simplifies the communication between two classes. In other words, it is just a contract, expressed as an interface.

Method: The command an agent exposes for responding to user needs.

Memento (Design Pattern): *See* Serialization.

Model-View-Viewmodel (Design Pattern): Separates the rendering of the view from the commands that control the view.

Mouse Jiggler: A software or hardware solution that causes the mouse to move by a tiny amount every second or so, to prevent the screen saver from

activating. Hardware mouse jigglers are USB devices that the system recognizes as a regular USB mouse.

Object: A concrete instance of a class or struct definition.

Polymorphism: The third pillar of Object Oriented Programming. Treat a child class as an instance of its base class, allowing the child class to override the base class implementation.

(Potentially) Shippable: Same as Shippable. 'Potentially Shippable' shouldn't be used to describe the state of a project. Either you have enough confidence in the product that you're willing to ship it to customers, or it needs work.

Property: The managed wrapper that controls access to the data encapsulated within a class or struct.

Prototype (Design Pattern): An object with a predefined state. Similar to an enumeration, these objects can be collected in a library where a predefined set of objects are useful.

Proxy (Design Pattern): A proxy is an agent that mediates communication between two clients.

Prototype Collection: A collection of prototypes that has special significance. See Enumeration.

Refactoring: The method used to change the implementation of a piece of functionality, without changing the way it is invoked.

Reference Type Object: A shared object that multiple agents have access to. Assignment operators normally copy the reference to the object and not the object.

Serialization: The process of capturing and restoring the internal state of an object.

Shippable: A product is shippable if you have enough confidence in it that, you're willing to bet your reputation on the product and give it to the customers.

Singleton: A class definition that only allows one instance of the agent.

Stakeholders: All people involved in the life-cycle of the product. This includes customers, investors, managers, and developers.

Strategy (Design Pattern): *See* Method.

Struct: The language construct used to define the properties and methods of a value type object.

Template: A template is a design pattern that allows us to abstract away the implementation details of a strategy.

Test case: A functional requirement, given form as verification code. Test cases are implemented in a library that is usually run during after each build.

Value Type Object: A non-sharable object which can have only one reference to it. Assignment operators can only copy the data

11. RESOURCES

Anything out there is vulnerable to attack
given enough time and resources.
-- Kevin Mitnick --

The internet is full of resources for developing solutions for helping our customers. Here are just some resources.

Remember to look at the terms of use for all online resources.

11.1. Sample Code and Blog
Here is some of my published work.

Sample Code
- Source: https://github.com/TrevyBurgess

UWP App - Zee Events Manager
- https://www.microsoft.com/en-us/p/zee-events-manager/9nblggh4vb78

Blog
- http://trevy-korner.blogspot.com/

11.2. Software Development Books
Here are some books in my library I would like to share with you.

Feature Management
1) Ed Tittel, Et Al, *.MCSD Analyzing Requirements and Defining .NET Solution Architectures Exam Cram 2 (Exam 70-300)*, May 2003
2) Robert E. Kelley, *How to Be a Star at Work: 9 Breakthrough Strategies You Need to Succeed*, June 1998
3) Ken Schwaber, *Agile Feature Management with Scrum*, March 2004

Software Development
1) Siobhán Clarke, Elisa Baniassad, *Aspect-Oriented Analysis and Design: The Theme Approach*, March 2005
2) Joseph Albahari, Ben Albahari, *C# 5.0 in a Nutshell: The Definitive Reference*, June 2012
3) Alan Shalloway, James R. Trott, *Design Patterns Explained: A New Perspective on Object-Oriented Design (2nd Edition)*, October 2004

11.3. Peer 2 Peer Assistance

The great power of the Internet is the ability to come together and help out each other with various challenges.

Web sites exist to ask questions of each other and share code. Here are some resources.

Programming assistance
- https://www.codeproject.com/
- https://stackexchange.com/
- https://stackoverflow.com/

Sample Code
- http://code.msdn.microsoft.com/

Open Source Project Hosting
- https://github.com

11.4. Open Source Resources

All good programs need media to help get a message to our customers. Also, images are a powerful method to communicate.

In software development, there is a trend for minimalist user interfaces. In my opinion, this makes it harder for users.

Easy to recognize icons can improve usability, when done correctly. Here is some useful media.

Clipart
- https://pixabay.com/
- https://classroomclipart.com/

Images
- http://OpenClipart.org/
- https://www.freeimages.com/
- http://clker.com/

Icons
- http://iconarchive.com/

11.5. Tutorials
- http://w3schools.com/

- http://wpftutorial.net/
- http://learnwpf.com/
- http://www.wpfsharp.com/

11.6. **Integrated Development Environments**

Desktop IDEs
1. http://eclipse.org/
2. https://netbeans.org/
3. http://www.icsharpcode.net/
4. http://sourceforge.net/
5. https://visualstudio.microsoft.com/

Inline IDEs
- https://dotnetfiddle.net/

11.7. **General Resources**

General
1. Big O Notation: http://BigOCheatSheet.com/
2. Connection Strings: http://www.connectionstrings.com/access-
    2007

12. INDEX

www.ingramcontent.com/pod-product-compliance
Lightning Source LLC
Chambersburg PA
CBHW030940180526
45163CB00002B/641